LIGHT AND ELECTRON MICROSCOPIC STUDIES OF ASCOBOLUS STERCORARIUS

II. Ascus and Ascospore Ontogeny

BY

KENNETH WELLS

UNIVERSITY OF CALIFORNIA PRESS
BERKELEY • LOS ANGELES • LONDON
1972

University of California Publications in Botany

Volume 62

Approved for publication September 15, 1971
Issued April 28, 1972

University of California Press
Berkeley and Los Angeles
California

University of California Press, Ltd.
London, England

ISBN: 0-520-09420-4
Library of Congress Catalog No.: 78-182553

printed in the united states of america

CONTENTS

LIGHT AND ELECTRON MICROSCOPIC STUDIES OF ASCOBOLUS STERCORARIUS

II. ASCUS AND ASCOSPORE ONTOGENY

BY

KENNETH WELLS

INTRODUCTION

THREE BASIC TECHNIQUES have been employed in the study of ascus and ascospore ontogeny. The earliest observations were of whole mounts of living or preserved specimens, or of sections, of ascocarps fixed usually in a solution of various proportions of chromic, acetic, and osmic acids. In most of these studies, staining was with Fleming's triple stain (i.e., safranin, crystal violet, and orange-G) or with haematoxylin and an appropriate counterstain.

Because of the intense interest in the nucleus in the early 1900's, especially in chromosome morphology during divisions, techniques were devised that enhanced the staining of the chromosomes but which distorted the remaining portions of the protoplasts of the ascus. These techniques usually utilized an alcohol-acetic acid fixative, after which the specimen was squashed in a carmine or an orcein stain. Such techniques resulted in great improvements in the knowledge of changes in chromosome morphology during meiosis and mitosis in the ascus (Olive, 1953) but were generally unsatisfactory for studies of other aspects of ascus development.

The most recent refinements in the techniques utilized in the study of ascus development have involved the electron microscope. The use of $KMnO_4$ as a fixative permitted an understanding of ascospore delimitation and development; however, only since the introduction of aldehydes and OsO_4 as the fixatives has it been possible to follow nuclear ultrastructure during ascus development and to correlate extranuclear changes with the nuclear cycle.

According to Buller (1915), Micheli (1729) first described and illustrated the ascus. He sketched and described the asci of several species of lichen-forming Ascomycetes and of several species of the Tuberales.

More detailed accounts of ascus development were published by de Bary (1863, 1866). In *Pyronema omphalodes* (Bulliard ex St. Amans) Fuckel [=*Pyronema confluens* (Pers. ex Pers.) Tul.] de Bary noted that the dense protoplast of the young ascus had relatively few vacuoles. During enlargement of the ascus, the basal three-fourths of the ascus protoplast became vacuolated. Smaller vacuoles also appeared in the distal region of the ascus. In the cytoplasm situated between these apical and basal vacuoles, three nuclear divisions occurred. Following the delimitation and the rearrangement of the ascospores, the epiplasm was characterized by its highly refringent and homogenous cytoplasm and by the presence of relatively large amounts of glycogen that stained reddish-brown with iodine. De Bary (1863, 1866) also studied ascus development in a number of other species.

Subsequently, Dangeard (1894) described the development of the crozier and the origin of the ascus from the penultimate cell of the crozier in *Peziza vesiculosa* Bulliard ex St. Amans and several other species. Further studies of crozier and ascus development were made by Claussen (1907, 1912) in *Pyronema omphalodes* (= *P. confluens*). Claussen noted that a crozier could originate from a segment of an ascogenous hypha, from a penultimate cell, or from a dikaryotic hyphal segment resulting from the fusion of a stalk and a terminal cell of a crozier.

Harper (1899) examined sections of the apothecia of *Scutellinia scutellata* (L. ex St. Amans Lambotte [= *Lachnea scutellata* (L. ex St. Amans) Gellet] and noted that the basal two-thirds of diploid asci and of more mature stages lacking differentiated ascospores were almost completely vacuolated. At these stages the apex of the ascus consisted of granular, essentially nonvacuolated protoplasm. In that portion of the protoplast between the granular apical zone and the nucleus or nuclei, there was an oval zone of relatively small vacuoles. Harper (1899) noted that the ascus protoplast in the vicinity of the nucleus or nuclei included few vacuoles.

The species of the genus *Ascobolus* have been the objects of a number of cytological studies (van Brummelen, 1967). A number of observations made during studies of *Ascobolus furfuraceus* Pers. ex Hook. and *A. stercorarius* (Bulliard ex St. Amans) Schroeter are relevant to the present study.

Harper (1896), Janczewski (1871), Schweizer (1932), and Welsford (1907) described the ascogonial coil in *A. furfuraceus* and noted that one segment became the ascogonium, which gave rise to the ascogenous hyphae. Janczewski (1871), studying *A. furfuraceus,* and Björling (1941), studying *A. stercorarius,* noted that the asci originated from the apices of the ascogenous hyphae; but they did not describe crozier formation. Crozier formation in *A. furfuraceus* was described by Schweizer (1932), and in *A. stercorarius* by Olive (1953) and Wells (1970). Vacuolation of the ascus protoplast in *A. furfuraceus* was noted by Janczewski (1871) and Fraser and Brooks (1909), and in *A. stercorarius* by Björling (1941).

Janczewski (1871) presented an accurate account of ascospore development. He concluded that the mature ascospore consisted of the protoplasmic contents (= sporoplasm), the colorless two-layered endosporium, the violet exosporium, and a jelly-like layer covering one side of the spore. According to the terminology adopted in this report, the two-layered endosporium described by Janczewski (1871) was probably the endospore and epispore, and his exosporium and jelly-like layer most likely corresponded to the perisporal sac.

Several theories have been proposed to account for "free cell" formation of ascospores and their subsequent development. These have been reviewed by Reeves (1967). Two postulates, however, have been especially prominent in the treatment of ascus ontogeny. The one proposed by Harper (1895, 1897, 1899, and especially 1905) was probably the most widely accepted explanation of ascosporogenesis prior to electron microscopic studies of ascus development (Olive, 1953). Harper (1895, 1897, 1899, 1905) expanded on Gjurasin's (1893) study of nuclear divisions in the asci of *Peziza vesiculosa.* Gjurasin (1893) was able to detect astral rays projecting outward from the poles of the dividing nucleus. Harper (1895, 1897, 1899, 1905) postulated that the astral rays arising from the centriolar plaque (= central body), which was located at the apex of the pear-shaped nucleus at the termination of division III, recurved around the nucleus. The astral rays, according to Harper, fused laterally to form a membrane separating the protoplast of the ascospore from the epiplasm. Harper (1895, 1897, 1899, 1905) proposed that as the ascospore matured the wall of the ascospore was deposited around the delimiting membrane, which became the plasma membrane of the ascospore.

Another hypothesis to explain the differentiation of the ascospores has become of importance since the electron microscope has been utilized in the study of ascus ontogeny. It apparently originated with some of the observations made by Mittmann (1933) in her study of *Ceratocystis fimbriata* Ell. & Halst. [≡*Ceratostomella fimbriata* (Ell. & Halst.) Elliot]. Mittmann noted that during the interphase between divisions II and III a central protoplasmic mass, including the four nuclei, was differentiated. She was not certain whether the central mass was differentiated by a wall, or whether it was a cavity; but she was able to detect narrow, unstained stripes around the periphery of the central mass. Mittmann (1933) reported that the ascospores were differentiated within the central zone.

Andrus and Harter (1933) examined sectioned material of *C. fimbriata* and concluded that the nuclear envelope of the fusion nucleus formed a vesicle in which meiosis and the first, post-meiotic, mitotic division occurred. After the delimitation of the ascospores by a process not described, the vesicle was ruptured. Thus, according to Andrus and Harter (1933), the nuclear envelope of the fusion nucleus became the wall of the ascus. In a more detailed study of *C. fimbriata* and *C. moniliformis* (Hedgc.) C. Moreau (=*Ceratostomella moniliformis* Hedgc.), Andrus and Harter (1937) again described and illustrated the ascus vesicle that was derived from the nuclear envelope. The vesicle in *C. fimbriata* was usually delimited by a wall, but in *C. moniliformis* the vesicle was defined by a cleavage space. In some cases, ascospore delimitation in *C. fimbriata* occurred within the walled ascus vesicle. In those asci of *C. fimbriata* with an unwalled vesicle, and in all asci of *C. moniliformis,* ascospore delimitation entailed an outward growth of the ascospores into the region of the ascus vesicle. Thus the concept proposed by Andrus and Harter (1937) was that the vesicle might, in some asci, function passively in ascospore delimitation.

Chadefaud (1943), studying *Mniaecia jungermanniae* (Fries) Boudier, noted a thin membrane surrounding the eight haploid spores. He did not believe that the delimited mass was a vacuole, as had been suggested by Gwynne-Vaughan and Broadhead (1936) in their study of *C. fimbriata,* since the delimited mass in *M. jungermanniae* included lipid globules and did not take up neutral red. Chadefaud (1943, 1960) suggested that the ascospores were delimited by progressive, inward cleavage of the enveloping membrane; hence, the role of the enveloping membrane in ascospore delimitation was an active one.

The first detailed report of ascus development observed with the electron microscope seems to have been that by Hashimoto, Gerhardt, Conti, and Naylor (1960). These authors examined thin sections of *Saccharomyces cerevisiae* Meyen fixed with $KMnO_4$. They reported that a membrane began to form around the individual nuclei even prior to the completion of meiosis. This became the plasma membrane of the ascospore. The outer coat of the ascospore wall was deposited first, according to these workers; and the inner coat, which was less eletron-dense, was deposited during the terminal stages of ascospore development. On the basis of present information derived from techniques and instruments more refined than those available to Hashimoto et al. (1960), the structure termed by these authors as the "outer coat" of the ascospore wall was, in some cases, the outer membrane of the double ascospore-delimiting membranes.

Conti and Naylor (1960) studied $KMnO_4$-fixed asci of *Schizosaccharomyces octosporus* Beijer. and reported that sporulation was initiated by the formation of two delimiting membranes around each nucleus. They suggested that the membranes might be initially connected. They also reported that the two-layered ascospore wall was deposited outside the delimiting membranes.

Marquardt (1963) examined sporulating cells of *Saccharomyces cerevisiae* fixed in $KMnO_4$ and found a double membrane surrounding each nucleus and a portion of the cytoplasm. Wall material was deposited between the two membranes. He suggested that the outer delimiting membrane became thicker and more electron-dense to form the outer spore layer. This concept was supported by Lynn and Magee (1970).

Other studies have extended the concept that the ascospore wall is deposited between a pair of ascospore-delimiting membranes. Although the number of these studies is limited, species of diverse taxonomic affiliations have been examined. The double ascospore-delimiting membranes have been detected in *Talaromyces vermiculatus* (Dang.) C. R. Benjamin (≡*Penicillium vermiculatum* Dang.) by Wilsenach and Kessel (1965); in *Cordyceps militaris* (L. ex St. Amans) Link by Moore (1965); in *Hansenula anomala* (Hansen) H. & P. Syd. by Bandoni, Bisalputra, and Bisalputra (1967); in *Podospora anserina* (Cesati) Niessl by Beckett, Barton, and Wilson (1968); in *Hypoxylon fragiforme* (Pers. ex Fries) Kickx. by Greenhalgh and Evans (1968); in *Hypoxylon rubiginosum* (Pers. ex Fries) Fries by Greenhalgh and Griffiths (1970); in *Sordaria fimicola* (Roberge) Cesati by Furtado and Olive (1970); in *Xylosphaera polymorpha* (Pers. ex Mérat) Dumortier by Beckett and Crawford (1970) and Schrantz (1970*b*); and in *Podospora setosa* (Winter) Niessl and *P. anserina* by Zickler (1970).

Among the studies of taxa of the Pezizales in which the ascospores have been shown to be delineated by the double ascospore-delimiting membranes are those by Delay (1966) of *Ascobolus immersus* Pers.; by Schrantz (1966, 1970*b*) of *Pustularia cupularis* (L. ex Fries) Fuckel and *Galactinia plebeia* Le Gal; by Bracker and Williams (1966) and by Bracker (1967) of *Ascodesmis nigricans* van Tieghem; by Carroll (1967*a*, 1969) of *Saccobolus glaber* (Pers. ex Pers.) Lamb [=*Saccobolus kerverni* (Crouan) Boudier]; by Reeves (1967) of *Pyronema domesticum* (Sowerby ex Gray) Saccardo; by Oso (1969) of *Ascobolus viridulus* Phill. & Plowr. (=*Ascobolus crenulatus* P. Karst.); and by Zickler (1970) of *A. immersus* and *A. stercorarius*.

Several investigators have demonstrated that the double ascospore-delimiting membranes are formed initially along the periphery of the ascus protoplast, and that the ascospores are differentiated by an inward growth or invagination of the delimiting membranes (Bracker, 1967, 1969; Carroll, 1967*a*, 1969; Greenhalgh and Griffiths, 1970; Oso, 1969; Reeves, 1967). These findings are in agreement with the concept proposed by Chadefaud (1943, 1960) based on his light microscopic studies of ascospore delimitation in *Mniaecia jungermanniae*.

Delay (1966) examined with both light and electron microscopy two strains of

Ascobolus immersus, one a wild-type strain with pigmented ascospores and the other a mutant with nonpigmented spores. She utilized both OsO_4 and $KMnO_4$ as fixatives and compared the images thus obtained. Since this study presents the most complete account of ascospore development from the delimitation stage to near maturity and since it is a study of a species closely related to *A. stercorarius,* the results are especially pertinent to the present report and have been most useful in interpretations of the electron micrographs of *A. stercorarius.*

Delay (1966) found that the ascospores were delimited by two membranes each having the characteristics of the plasma membrane. The electron-transparent endospore was formed between the two ascospore-delimiting membranes. The inner ascospore-delimiting membrane became the plasma membrane of the spore. The stratified epispore was formed outside the endospore prior to maturity. Although initiated before the epispore, the mucilaginous perisporal sac was formed at essentially the same stage. Subsequently, the pigment granules appeared in the perisporal sac of the ascospores of the wild-type strain. Such granules were lacking in the mutant strain forming nonpigmented ascospores. Thus the pigment of the wild-type spores was restricted to the perisporal sac.

At the time of the delimitation of the ascospores of *A. immersus,* according to Delay (1966), the ultrastructural characteristics of the sporoplasm and the epiplasm were identical, except that organelles and membranes were more abundant in the epiplasm. The changes in the ultrastructural features of both the sporoplasm and the epiplasm during ascus maturation were described in detail by Delay. She noted a characteristic and regular orientation of the cisternae of the endoplasmic reticulum in the epiplasm adjacent to the investing membrane during the development of the endospore. The wall of the ascus of *A. immersus* consisted of a broad, electron-transparent inner layer and a narrow, electron-dense outer layer.

A study of *A. stercorarius* has been made consisting of observations of smear preparations, of thick sections of apothecia fixed in acrolein and embedded in glycol methacrylate, and of thin sections of apothecia fixed in $KMnO_4$ or in glutaraldehyde and OsO_4. In addition, phase-contrast and interference-contrast optics were utilized to examine living material. A report on the observations bearing upon the three nuclear divisions in the ascus of *A. stercorarius* has been made (Wells, 1970). In the present report, the extranuclear changes that occur in the ascus and ascospores during their ontogeny are described. The growth and development of an ascus is a dynamic process that cannot be conveniently divided into distinct steps that may be described as separate and unrelated processes. Obviously, the nuclear cycle and the events described in the present report are closely interdependent; therefore, I have made frequent references to the nuclear cycle. It is most unfortunate that available techniques do not permit studies of ascus development *in vivo;* however, the development sequence described herein is substantiated by my own observations and those of others reported in the available literature.

Acknowledgments

I wish to thank Dr. Ching-Yuan Hung for his most skillful technical assistance throughout this study. Gratitude is also due Professor T. Elliot Weier and Mr. E. B. Risley for the many courtesies they have extended and Professor G. C. Carroll for his most helpful review of the manuscript. This study would not have been possible without the financial support of Research Grants GB-517 and GB-12929 from the National Science Foundation. I also wish to thank the management of *Mycologia* for permission to reproduce the electron micrographs of plates 13 and 15, which were used as illustrations in a previous report (Wells, 1970).

TERMINOLOGY AND NOMENCLATURE

A perusal of the literature pertaining to ascus development reveals the need to clarify some of the terminology required in discussions of this subject. Several of the terms applied in recent studies utilizing the electron microscope have not as yet become stabilized; therefore, it is desirable to outline here my interpretations of the terms.

As in an earlier report (1970), I have followed the terminology applied to the segmented crozier by Wilson (1952) and by Beckett and Wilson (1968). Thus the *terminal cell* is the distal, uninucleate hyphal segment, the *penultimate cell* is the subterminal, curved, dikaryotic hyphal segment, and the *stalk cell* is the basal, uninucleate hyphal segment.

The *sporogenous zone* of the ascus refers to the portion containing the ascospores or to the portion in which the ascospores will be delimited. Prior to ascospore delimitation, the sporogenous zone in most asci is delineated apically by one to several clusters of apical vacuoles and basally by the large vacuoles that occupy the basal half to two-thirds of the ascus.

Following delimitation of the ascospores, that portion of the cytoplasm, including vacuoles, not enclosed within the ascospores is referred to as the *epiplasm* (de Bary, 1863). That portion of the cytoplasm enclosed within the ascospore is designated as the *sporoplasm* (Delay, 1966).

Shortly after the diploid nucleus is formed, portions of the ascus protoplast of *Ascobolus stercorarius* become vacuolated. *Apical vacuoles* refer to the one to several clusters of small vacuoles above the nucleus or nuclei, and the *basal vacuoles* refer to the vacuoles beneath the nucleus or nuclei.

Nuclear bleb is used in the sense of Carroll (1967*a*), i.e., a small vesicle or sac that theoretically originates by evagination from the nuclear envelope and an associated segment of endoplasmic reticulum. The *double ascospore-delimiting membranes* are the membranes that form along the periphery of the sporogenous zone and subsequently invaginate, or develop inwardly, to differentiate the ascospores. Following the complete delimitation of the ascospores, the inner double ascospore-delimiting membrane becomes the plasma membrane of the ascospore, whereas the outer delimiting membrane becomes the *investing membrane* (Moore, 1963; Carroll, 1967*a*, 1969; Reeves, 1967). At all stages of maturation that could be satisfactorily fixed and sectioned, the investing membrane defines the outer boundary of the ascospore.

The terms applied to the layers of the ascospore wall follow the definitions of Chadefaud (1960) and Delay (1966). Thus, the *endospore* is the inner layer, which in *A. stercorarius* is electron-transparent. The *epispore* is the outer layer of the ascospore wall proper. The *perisporal sac* is the pigmented, mucilaginous layer surrounding the endospore and the epispore.

The *dikaryotic* nuclear stage refers to the binucleate state in the ascogenous hyphae, penultimate cell, and ascus prior to karyogamy. *Two-nucleate* is used to describe the nuclear stage following division I, *four-nucleate* indicates the nuclear stage following division II, and *eight-nucleate* is the nuclear condition following division III.

To avoid the excessive use of descriptive terms, I have not considered *vacuoles* as part of the cytoplasm of the ascus or of the ascospores. The *ascus protoplast* is considered as all portions of the ascus within the ascus wall and outside the investing membranes of the delimited ascospores. The *ascospore protoplast* is defined as all portions of the spore within the endospore.

Where possible, the nomenclature proposed by Dennis (1968) and by van Brummelen (1967), except in the case of *A. furfuraceus,* is followed. According to van Brummelen (1967), *A. stercorarius* (Bulliard ex St. Amans) Schroeter is a later synonym of *A. furfuraceus* Pers. ex Hook.; however, Olive (1954) maintained that isolates of the two species were intersterile and were morphologically distinguishable. Because of Olive's findings and because the two epithets are often treated in the literature as representing different species, I have treated the two names as representing distinct taxa. It is not my purpose, however, to express a taxonomic judgment on this matter.

MATERIALS AND METHODS

The strain of *A. stercorarius* (Bulliard ex St. Amans) Schroeter used and the culture methods were as described earlier (Wells, 1970). In addition to malt-yeast extract agar, yeast-starch agar (Emerson, 1958) was used for the production of apothecia.

For light microscopy, developing apothecia were fixed in 10 percent acrolein and embedded in glycol methacrylate following the techniques described by Feder and O'Brien (1968). Sections 1 to 4μ in thickness were cut with glass knives and stained with a variety of dyes; however, toluidine blue (Feder and O'Brien, 1968) produced the most satisfactory results, especially for photomicrography. Some observations were made of material fixed in carnoy's solution and stained with acetocarmine (Wells, 1970).

For observations of living material, thick sections of apothecia, cut parallel to the long axes of the asci, were placed in distilled water on a glass slide and gently squashed beneath a coverslip. These slides were examined with Zeiss phase-contrast and Zeiss Nomarski's interference-contrast optics. The protoplasts of asci prior to the delimitation of ascospores became distorted within 3 to 5 minutes, and all efforts to find a more suitable mounting medium for observing the ascus in the living state were unsuccessful.

The OsO_4 and Sudan black B techniques, following the procedures as described by Jensen (1962), were used as tests for lipids. The Bauer-Feulgen and Best carmine techniques (Glick, 1949) were used in an effort to detect glycogen at various stages of ascus development.

For electron microscopy, apothecia were fixed in 1.5 percent $KMnO_4$, dehydrated in an acetone series, and embedded in Maraglas. Other material was fixed in 1 to 6.5 percent glutaraldehyde buffered at 7.0 to 7.4 with 0.15 M phosphate at 0 to 4 C for 30 minutes to 4 hours. The material was postfixed in 1 to 2 percent OsO_4 in the same buffer for 1 to 2 hours at 0 to 4 C. The glutaraldehyde-OsO_4-fixed material was also dehydrated in an acetone series and embedded in Maraglas. The thin sections were stained with various combinations of lead citrate, barium permanganate, and uranyl acetate. A Hitchi HU-11 electron microscope was used for all electron micrographs.

All magnifications and measurements given are only approximations.

OBSERVATIONS

GENERAL

The distinct hymenium of paraphyses and asci is evident in apothecia sectioned parallel to the long axes of the asci (pl. 1, A). In such sections of apothecia with only a few mature asci, all stages of ascus development are visible (pl. 1, A). The characteristic feature of rapid enlargement of the ascus in the genus *Ascobolus* (van Brummelen, 1967), especially following delimitation of the ascospores, is evident in these sections (pl. 1, A). To further demonstrate this feature, 20 asci at the stages of development indicated in table 1 were measured with phase-contrast optics. The results indicate that the length and width of the asci double even after a portion of the endospore has been deposited.

The hymenium is covered by a layer of segmented paraphyses (pl. 1, A) which is penetrated only by the maturing asci immediately prior to ascospore discharge (Corner, 1929; Gamundí and Ranalli, 1963). The hyphal elements of all portions of an apothecium are embedded in a gelatinous matrix (Corner, 1929; Gamundí and Ranalli, 1963; Janczewski, 1871). Electron-dense areas of this matrix adjacent to the asci and paraphyses are often visible in the electron micrographs of apothecia fixed in $KMnO_4$ (pls. 4, A; 7, A; 9; 18).

CROZIER AND DIKARYOTIC ASCUS STAGES

As in many other species of the Euascomycetidae (Carroll, 1969; Hung and Wells, 1971; Olive, 1953; Reeves, 1967), the ascus in *Ascobolus stercorarius* is derived from the penultimate cell of a crozier (Olive, 1953; Wells, 1970). The ultrastructural characteristics of a developing (pl. 4, A) or segmented crozier (Wells, 1970) are essentially identical with those of a dikaryotic ascus (pl. 4, B). At both stages there are one to several small, basal vacuoles (pl. 4, A). In $KMnO_4$-fixed apothecia, the tonoplasts of the basal vacuoles, especially in the earlier stages of ontogeny, are often discontinuous (pl. 4, A). Since this feature is not evident in glutaralydehyde-OsO_4-fixed apothecia, these discontinuities are considered to be artifacts induced by $KMnO_4$ fixation. The mitochondria at these early stages are usually long, sinuous, and sometimes branched (pl. 4, A, B). Elements of the endoplasmic reticulum are usually conspicuous throughout the cytoplasm and often form a netted pattern in both the basal and apical portions of the ascus (pl. 4, A, B).

The remaining features of the cytoplasm of croziers and dikaryotic asci are essentially as described for the hyphal elements of the somatic hyphae of other filamentous Ascomycetes (Bracker, 1967). However, Woronin bodies were not found in the penultimate cell of the crozier or in the ascus, and were present only rarely in the ascogenous hyphae and in the stalk and terminal cells of segmented croziers. Woronin bodies do occur in the sterile elements of the apothecium, such as the paraphyses (pls. 13; 21, A). Also, the electron-dense masses adjacent to prefusion and fusion nuclei, which are preserved only by glutaraldehyde-OsO_4 fixative, are found only in the ascogenous hyphae, croziers, dikaryotic asci, and diploid asci (Wells, 1970; pl. 6, A). In apothecia fixed in $KMnO_4$ they appear as clear areas adjacent to the nuclear envelope (pls. 4, A, B; 5, A, B; 7, A, B).

TABLE 1
DIMENSIONS OF ASCI AT SEVERAL STAGES OF ONTOGENY

Stage of ontogeny	Range of sizes, μ	Average size, μ
Diploid asci with apical vacuoles (e.g., pl. 1, C, D)	41.5–95 × 9–12.5	71.2 × 10.9
Asci with spherical ascospores without an endospore (e.g., pl. 2, D)	82–120 × 9–12.5	95.2 × 10.9
Asci with ascospores with an endospore but without a perisporal sac (e.g., pl. 2, E)	92.5–138.5 × 10–14	106.2 × 12.5
Asci containing ascospores with nonpigmented perisporal sac (e.g., pls. 2, F; 3, a)	138–210 × 15–25	173.7 × 20.4
Mature asci (e.g., pl. 3, C)	164–271 × 21.5–33*	215.5 × 27.8

* Van Brummelen (1967) stated that the mature asci of *Ascobolus furfuraceus*, in which he includes *A. stercorarius*, are 180–250 × 24–30μ.

Diploid Ascus Stage

Karyogamy takes place at a relatively early stage in the development of an ascus. It occurs after the nuclei have assumed a vertical position with respect to one another (pl. 4, B). Stages of nuclear fusion have not been observed in the electron micrographs.

Within the ascus protoplast, several changes typically occur during the diploid stage. Small vacuoles develop above the nucleus, a zone essentially devoid of organelles and membranes forms in the dome at the tip of the ascus, and the basal portion of the ascus becomes progressively vacuolated. In addition, the pores at the base of the ascus become plugged, and mitochondria undergo a change in morphology.

A recently formed diploid ascus lacks a significant accumulation of vacuoles in the cytoplasm in the distal portion of the ascus (pls. 1, B; 5, A). Subsequently, a netted, endoplasmic reticulum system appears in this region. In the electron micrographs of gluteraldehyde-OsO_4-fixed apothecia, it can be demonstrated that this netted membrane system is smooth (pl. 6, A)—i.e., that the membranes are not associated with ribosomes as are the other elements of endoplasmic reticulum in the ascus (pl. 6, A). The granular endoplasmic reticulum is present as isolated strands and is closely associated with the smooth, netted elements. Continuities between the two types are frequent (pl. 6, A, B). The smooth, netted endoplasmic reticulum often forms an extensive system in the distal portion of the ascus (pls.

5, B; 6, B). Some densely stained regions above the diploid nucleus, visible in thick sections stained with toluidine blue, are interpreted as accumulations of netted endoplasmic reticulum (pl. 1, B). Such extensive accumulations of the netted endoplasmic reticulum are rarely seen beside or below the diploid nucleus after the appearance of apical vacuoles (pl. 7, A, B). The systems of netted endoplasmic reticulum noted in the distal region of the diploid ascus do not exhibit the regularity and compactness of structure of the lattice bodies described by Zachariah (1970) from the excipulum of the apothecium of *A. stercorarius.* I have detected lattice bodies in the ectal and medullary excipula of apothecia fixed in $KMnO_4$ and in glutaraldehyde-OsO_4. Lattice bodies are apparently not present in the ascogenous hyphae and asci of *A. stercorarius.*

As the diploid ascus continues to enlarge, small vacuoles appear in the distal region of the cytoplasm (pls. 1, C, D; 5, B; 7, A, B). Studies of a number of thin sections of diploid asci indicate that the membranes of the netted endoplasmic reticulum fragment to form flattened cisternae that then become inflated (pls. 5, B; 7, A, B). The apical vacuoles are generally more spherical in living material (pl. 1, D) than in apothecia fixed in $KMnO_4$ (pl. 7, B). In glutaraldehyde-OsO_4-fixed ascocarps, the apical vacuoles are spherical, similar to the vacuoles illustrated in plate 15. The individual apical vacuoles are often not visible in acrolein fixed apothecia (pl. 1, C).

During development of the apical vacuoles, a region at the tip of the ascus becomes nearly devoid of mitochondria and segments of the endoplasmic reticulum (pl. 7, B). Evidence of this apical zone can be seen in living asci (pl. 1, D) and in the thick sections stained with toluidine blue (pl. 1, B, C). The extent of this apical clear zone varies; therefore, it is certainly not a fixed area but one which constantly changes in extent and form. It is, however, a consistent feature of diploid asci with apical vacuoles. Within this apical clear zone are scattered small vesicles (pl. 7, B; see also the vesicles in the four-nucleate ascus of pl. 9) that are similar in structure to the apical vesicles described by Brenner and Carroll (1968) in the hyphae of *Ascodesmis sphaerospora* Obrist, by Girbardt (1969) and Grove and Bracker (1970) in the hyphal tips of a variety of species, and by Grove, Bracker, and Morré (1970) in the hyphal apices of *Pythium ultimum* Trow. I believe that the apical zone nearly devoid of organelles and membranes in the asci of *A. stercorarius* is associated with the apical growth of the ascus. Also, it seems probable that this apical zone is identical with the apical zone of "granular," nonvacuolated protoplasm noted by Harper (1899) in the asci of *Scutellinia scutellata.*

As the ascus enlarges rapidly during the diploid stage, the basal half to two-thirds of the protoplast becomes filled with large vacuoles, with only a narrow zone of cytoplasm remaining at the base and around the periphery of the ascus (pls. 1, B, C, D; 5, B; 7, A, C). Smaller vacuoles are visible in the upper regions of this vacuolated zone (pls. 1, C, D; 7, A), and it appears that the vacuoles originate in this region (pl. 7, A). The uninflated membranes that become the tonoplasts of the basal vacuoles often form an inverted dome-shaped configuration in favorable thin sections of this portion of the ascus (pl. 7, A).

In the cytoplasm, which becomes concentrated in the vicinity of the diploid nucleus in the upper third to half of the ascus, a characteristic change can be detected in the shape of the mitochondria. Most of the mitochondria assume a curved, dumbbell-like shape in longitudinal section (pl. 7, B). In transverse sections of the semicircular mitochondria, the outline appears as a double circle, with the two membranes more widely separated on one or two sides (pl. 7, B). Delay (1966) also reported the presence of many mitochondrial profiles in the form of a dumbbell during ascus development in *Ascobolus immersus.*

During the diploid and subsequent stages of ascus development, stacks of parallel lamellae have been observed quite rarely (above the upper, right ascospore initial in pl. 16). However, the cisternae of these configurations in glutaraldehyde-OsO_4-fixed material are closely associated with ribosomes.

At the base of the ascus, the ascomycetous septa formed between the ascus and the stalk cell and between the ascus and terminal cell of the initial crozier increase in thickness (pls. 7, C; 8, A, B). For example, the basal septa in plate 4, A, average 0.12μ in thickness, whereas those in plate 8, B, average 0.29μ. The deposition of wall material seems to occur mainly on the ascus side of the delimiting septa. The pores, which are initially of uniform diameter in the crozier, become concave in outline, with the larger diameter of the pore adjacent to the ascus cytoplasm (pls. 7, C; 8, A, B). Thus, in my opinion, the narrow rims of the lower edges of the pores in plate 8, A, B, represent the original wall material of the simple septa formed during segmentation of the crozier. This is indicated by the fact that the narrowest diameters of the pores measured from several electron micrographs of segmented croziers and diploid asci are approximately equal, varying between 0.2μ and 0.26μ.

During the diploid stage the basal septa become plugged. In the initial stages of formation, the thick, convex plug is only slightly electron-opaque and is often closely associated with endoplasmic reticulum (pl. 8, A). Subsequently, the plugs in the basal septa become electron-dense and appear to be structurally capable of preventing the loss of turgor pressure from the ascus, and able to arrest the movement of any substances out of, or into, the ascus (pl. 8, B). The basal plugs are quite conspicuous in the basal septa of the asci stained in acetocarmine at the diploid or subsequent stages (pl. 1, E), and can, with difficulty, be detected in living material with the use of phase-contrast optics.

The electron-dense plugs in the pores of the basal septa of the asci of *A. stercorarius* are unlike the plugs in the basal septa of the asci of *Ascodesmis sphaerospora,* as described by Carroll (1967*b*), and *of Galactinia plebeia,* as described by Schrantz (1970*b*). The plugs in the basal septa of the asci of *Saccobolus glaber* (Carroll, 1967*b*) were closely associated with segments of the endoplasmic reticulum and appear structurally similar to the developing plugs in the basal septa of the asci of *A. stercorarius* (pl. 8, A). Hemispherical plugs have been detected with light microscopy in the basal septa of the asci of a variety of species (Chadefaud, 1960). Additional studies, with the electron microscope, are needed of the basal septa and associated pore plugs of the asci of a variety of species before any generalities can be made. It would appear, however, that there are structural and developmental differences between the plugs in the basal septa of the asci of diverse taxa.

Lomasomes (i.e., clusters of tubules or vesicles outside the plasma membrane) are present infrequently in the basal region of the ascus at the diploid (pl. 8, A) and subsequent stages. Lomasomes are very rarely present between the investing membrane and endospore (pl. 20), or between the plasma membrane of the sporoplasm and the endospore, during the early stages of ascospore maturation. Lomasomes are not consistently abundant at any particular stage of ontogeny, nor are they consistently associated with any structure or portion of the ascus in *A. stercorarius*. Thus it is not possible to ascribe a function to the lomasome in the ascus of *A. stercorarius*.

There are frequent vesicular invaginations of the plasma membrane of the ascus protoplast in the sporogenous zone of the ascus at various stages of ontogeny that I do not consider as representing lomasomes (pl. 29, A, B, C). Presumably, these configurations are formed by invaginations of the plasma membrane in a circular pattern and fusion of the inner margin of the invaginations. A variety of forms are present (pls. 29, A, B, C; 13). These structures seem to be most abundant during the delimitation of the ascospores (pl. 13); however, they are present at all stages of ontogeny. Possibly, they are artifacts, but they do occur in both $KMnO_4$- and glutaraldehyde-OsO_4-fixed material.

Two- and Four-nucleate Ascus Stages

Meiosis takes place resulting in the formation of four haploid nuclei (Wells, 1970). Other than the occurrence of this significant process, the changes that occur in the ascus involve mainly an increase in size associated with an apparent increase in the amount of cytoplasm adjacent to the nuclei (pls. 2, A, B; 9). However, the presence of vacuoles in the sporogenous zone makes it difficult to determine if there is a real increase in cytoplasm. The basal and apical vacuoles persist in the two- and four-nucleate asci (pl. 2, A, B) essentially as described for the diploid ascus. The ultrastructural features of the cytoplasm in the sporogenous zone (pl. 9) resemble those of the diploid ascus with apical vacuoles. In addition to the apical and basal vacuoles, portions of the cytoplasm in the peripheral regions of the ascus between the two or four nuclei often become vacuolated (pl. 2, A). The zone at the apex that is devoid of membranes and organelles is of variable extent, being present (pl. 9), much reduced, or possibly absent (pl. 2, A, B). When the distal clear zone is present, apical vesicles are evident in favorable electron micrographs (pl. 9).

Delimitation of Ascospores and Eight-nucleate Stage

As noted by Zickler (1970) in several ascomycete species, it appears that the formation of the double ascospore-delimiting membranes in *A. stercorarius* is initiated along the periphery of the ascus during the four-nucleate stage. This is indicated by the failure to observe an ascus in a stage earlier than the four-nucleate stage with the double ascospore-delimiting membranes and by the frequent observation of four-nucleate asci without (pls. 9; 10, A) and with (pl. 10, B) the ascospore-delimiting membranes. A few asci with nuclei apparently at prophase III possess the delimiting membranes (Wells, 1970). Therefore, I suggest that the delimiting membranes are formed along the periphery of the ascus during the interphase between division II and division III. It is possible that the initiation of this membrane system is not correlated precisely with the nuclear cycle, but evidence for this is lacking.

The evidence obtained in this study supports the hypothesis proposed by Carroll (1967*a*, 1969) that the double ascospore-delimiting membranes are derived from nuclear blebs. Images suggesting the formation of nuclear blebs are most abundant at the four-nucleate stage; however, similar images do occur infrequently in the vicinity of diploid nuclei and on nuclei in ascospores with a well-formed endospore. Segments of the endoplasmic reticulum become closely associated with the nuclear envelope (pl. 11, A). In sections of asci fixed in $KMnO_4$ and stained with lead citrate, the associated nuclear envelope and the segment of endoplasmic reticulum are more electron-dense than unassociated elements of these membranes (pl. 11, A, B). In addition, the distance between the two membranes of the associated nuclear envelope and the segment of endoplasmic reticulum in reduced (pl. 11, A, B; see also Carroll, 1967*a*). The associated membranes evaginate (pl. 11, B) to form a spherical to oval nuclear bleb (pls. 11, A; 12, A, B). The matrix of the nuclear bleb is, in some instances, more electron-dense than the cytoplasmic ground substance (pl. 12, A, B), but the difference in density, when evident, is slight.

As proposed by Carroll (1967*a*), the nuclear blebs appear to migrate to the periphery of the ascus and to unfold to form the double ascospore delimiting membranes (pl. 12, A, B). The delimiting membranes develop at several points along the ascus wall in the basal portion of the sporogenous region (pl. 10, B). Generally, the lumen between the two delimiting membranes has a constant width, and the distance of the two membranes from the ascus wall is essentially uniform (pl. 12, C). Rarely, the delimiting membranes form elaborate configurations (pl. 12, C). In a number of electron micrographs, the nuclei nearest the developing edges of the double ascospore-delimiting membrane are spatially, or directly, associated with the greatest number of nuclear blebs (pl. 10, B).

In favorable thin sections of glutaraldehyde-OsO_4-fixed apothecia, the plasma membrane of the ascus protoplast at the four-nucleate stage (pl. 28, A), as at all other stages of ascus ontogeny, has a distinct, asymmetrical, tripartite aspect. The broader, electron-dense component is adjacent to the ascus wall (pl. 28, A). Similarly, the individual membranes of the double ascospore-delimiting membranes have an asymmetrical, tripartite structure in favorable thin sections (pl. 28, A). The broader, electron-dense components of the two membranes are both situated toward the interior of the double membrane complex (pl. 28 A). Such an orientation of a double membrane complex of asymmetrical, tripartite membranes would result from the invagination of a single asymmetrical, tripartite membrane. Invaginations of the plasma membrane of the ascus protoplast have such an orientation of the broader, electron-dense components (pls. 28, A; 29, A, B). However, in all invaginations of the plasma membrane observed the interspace between the apposed broader, electron-dense components was less than the similar interspace in the delimiting membranes (pls. 28, A; 29, A, B, C). The invaginated plasma membrane appears more electron-dense than do the double ascospore-delimiting membranes (pls. 28, A; 29, C), possibly because of more tenuous interspace between the members of the invaginated plasma membrane as compared with the interspace between the two ascospore-delimiting membranes.

In material fixed in glutaralydehyde-OsO_4, there is an electron-dense layer in some regions between the double ascospore-delimiting membranes (pls. 28, A; 29,

C). This median layer is found in certain portions of the delimiting membranes during the four-nucleate stage (pls. 28, A; 29, C) and before and after the delimitation of the ascospores during the eight-nucleate stage. It was not detected in $KMnO_4$-fixed apothecia.

A number of electron micrographs of *A. stercorarius* indicate that during division III the double ascospore-delimiting membranes have the three-dimensional aspect of an open cylinder with discontinuous sides. As described by Carroll (1967*a*) for *Saccobolus glaber,* the delimiting membranes of *A. stercorarius* develop across the basal region of the sporogenous zone and across the distal portion, but below the apical vacuoles, of the sporogenous zone. Electron micrographs indicating that the delimiting membranes form a continuous sac have not been obtained; discontinuities in the membranes are always present (pl. 13).

At the completion of division III, the double ascospore-delimiting membranes develop inwardly, essentially as a single membrane system, between the eight haploid nuclei (pl. 13). Because of the frequent discontinuities in the delimiting membranes (pls. 13, 14, 15), the presence of nuclear blebs during the process of ascospore differentiation (pl. 14), and the resemblance of the sporoplasm and the epiplasm during and after ascospore delimitation (pls. 14, 16), I believe there is an inward growth of the delimiting membranes, possibly accompanied by some invagination. No evidence was obtained that the membranes and organelles are forcibly swept into the differentiating ascospores (pls. 14, 16, 18). Although the astral rays radiating from the centriolar plaque seem to be more abundant at the termination of division III than at other stages of nuclear divisions in the ascus (Beckett and Crawford, 1970; Wells, 1970; Zickler, 1970), there is no convincing evidence to signify that the astral rays function in ascospore delimitation. The presence of numerous astral rays at this stage, however, does support the plausibility of the suggestions made by Carroll (see Bracker, 1967), Bracker (1967), and Beckett and Crawford (1970) that the astral rays function as barriers, or as active agents, in the displacement of the double ascospore-delimiting membranes.

Apparently the ascospore-delimiting membranes separate into eight parts and surround the individual nuclei delimiting the ascospores (pl. 14). Logically, the separations would occur in those regions of the delimiting membranes nearest the axis of the ascus. This is indicated by the fact that in the terminal stages of ascospore delimitation, the discontinuities are present in the delimiting membranes toward the center of the ascus (pls. 14, 15). Nuclear blebs are often visible within the cytoplasm partially enclosed by the delimiting membranes (pl. 14) and are even in the sporoplasm of seemingly completely delimited ascospores (pl. 16).

At the time of delimitation, the ascospores are arranged in four pairs and are somewhat flattened in outline (pls. 14, 15, 16). The spindles of division III in *A. stercorarius* are essentially transverse, but the daughter nuclei soon assume a slightly oblique orientation (pl. 15; Wells, 1970). Thus the pairs of recently delimited spores have a slight oblique orientation (pls. 14, 15, 16). This orientation is altered after delimitation in such a way that the ascospores form an alternating pattern and, therefore, occupy a larger part of the ascus (pls. 2, D; 18; 19).

The apical vacuoles, which persist in the distal portion of the ascus until the ascospores are differentiated (pls. 14, 16), disappear from the tip of the ascus at the time of reorientation of the ascospores (pl. 18). Small vacuoles appear within the epiplasm (pls. 15, 18); and a portion of these, I believe, are displaced apical vacuoles. During the eight-nucleate stage and the terminal stages of ascospores delimitation, regions of the cytoplasm of the sporogenous zone become vacuolated (pls. 2, C; 18). These latter vacuoles may also become the interascospore vacuoles. The apical zone devoid of membranes and organelles, which is most conspicuous in diploid asci (pls. 1, C, D; 7, A, B), is reduced; but apical vesicles are often present in the tip after ascospore delimitation (pl. 18) and at subsequent stages (pl. 19).

Except for the increase in concentration of glycogen granules, the ultrastructural characteristics of the ascus protoplast at this stage of ontogeny (pls. 13, 14, 15, 16, 17, 18) remain essentially as in the diploid ascus.

Positive tests for glycogen following the Bauer-Feulgen staining procedure (Glick, 1949) were obtained in the epiplasm of asci with delimited ascospores. The most intense staining by this technique occurred in the peripheral cytoplasm of asci which had ascospores with unilateral perisporal sacs, especially in the cytoplasm in the distal region of the ascus adjacent to the operculum. The results obtained with the Best carmine technique, as described by Glick (1949), were less explicit than those obtained with the Bauer-Feulgen procedure. Glycogen could not be demonstrated by either technique in mature asci prior to ascospore discharge, or in the protoplasts of asci following ascospore release.

In the electron micrographs, glycogen granules appear sparsely in the protoplasts of diploid asci (pls. 8, B; 28, C). Such granules become increasingly more abundant in four-nucleate asci (pls. 10, A; 12, A; 29, C), eight-nucleate asci (pls. 13; 14; 15; 28, D), and in the epiplasm of asci with delimited ascospores (pls. 16; 17; 18; 21, B; 22, B; 23; 24, A, B).

In diploid and four- and eight-nucleate asci fixed in glutaraldehyde and OsO_4, glycogen granules often appear at the periphery or in the outer regions of concentric membrane complexes (pls. 8, B; 15; 28, C, D). Some of the glycogen granules, seemingly unassociated with concentric membrane complexes in the electron micrographs of asci at these stages, can be shown to be adjoining concentric membrane complexes in adjacent thin sections. Because of the technical problems involved in preparing and examining serial thin sections, it is, of course, not possible to demonstrate that all glycogen granules are associated with a concentric membrane complex in diploid and four-and eight-nucleate asci (e.g., pl. 17). In subsequent stages of ascus ontogeny in apothecia fixed in glutaraldehyde-OsO_4, glycogen granules become more concentrated in the peripheral portions of the epiplasm and usually do not appear to be associated with concentric membrane complexes (pls. 22, B; 24, B; 25, C).

The outer membranes of the concentric membrane complexes are often continuous with the plasma membrane (pl. 28, A). These membrane complexes are composed of annular pairs of membranes that are uniformly spaced (pl. 28, B). The pairs vary in their spatial relationship with other pairs composing the complex (pl. 28, B). The membranes composing the outer pairs have the asymmetrical, tripartite structure of the plasma membrane and of the double ascospore-delimiting membranes (pl. 28, A, B, D). In some electron micrographs, the asymmetrical structure of the membranes composing the concentric membrane complex, the ascospore-delimiting membranes, and the plasma membranes is visible (pl. 28, A, B, C, D). In the concentric membrane complex, the more electron-dense layer of the individual tripartite membranes is situated adjacent to the uniform space between the pairs of tripartite membranes (pl. 28, A, B, C, D). The plasma membrane and the membranes of the concentric membrane complexes are somewhat more electron-dense than the ascospore-delimiting membranes (pl. 28, A). The inner pairs of membranes of the concentric membrane complex are less electron-dense than the peripheral pairs and often become indistinct (pls. 17; 28, A, C, D).

It seems possible that the concentric membrane complexes are formed by invagination of the plasma membrane of the ascus. Initial stages of formation may be similar to the invaginations illustrated in plate 29, A. Possibly, additional coils are formed within the invaginations; however, the exact mechanism cannot be followed by electron microscopy. There is a range of structures similar to those illustrated in plate 29, A, and in plate 28. Thus, the similarities in membrane morphology, and the continuities observed (pl. 28, A), indicate that the concentric membrane complexes associated with the formation of glycogen granules may be derived from the plasma membrane of the ascus protoplast. If this is a correct postulate, the membranes forming the concentric membrane complexes must undergo an alteration in structure, since the concentric membranes are not preserved in $KMnO_4$ fixed apothecia (pls. 10, A, B; 14; 16; 18).

Ascospore Ontogeny

At the time of complete differentiation from the epiplasm, the ascospore is surrounded by two membranes and is essentially spherical to slightly flattened in median sections (pls. 2, D; 17; 18). The membranes surrounding some ascospores fixed in $KMnO_4$ are interrupted (pl. 18); however, since such breaks do not occur in glutaraldehyde-OsO_4-fixed materials (pl. 17), these discontinuities are considered to be artifacts of $KMnO_4$ fixation. Except for the absence of vacuoles, the ultrastructural features of the sporoplasm in the early stages of ascospore development (pls. 14, 15, 16, 17, 18) are as described for the cytoplasm of the diploid ascus. I have no logical explanation for the fact that the sporoplasm of ascospore initials fixed in glutaraldehyde-OsO_4 is less electron-dense than the adjacent epiplasm (pl. 17), unless the difference is related to the penetration of the fixatives. This difference is not apparent at earlier stages (pl. 15) or after the endospore layer of the wall is initiated (pl. 22, B).

The electron-transparent endospore is deposited between the two delimiting membranes (pl. 19). Thus, the inner delimiting membrane becomes the plasma membrane of the sporoplasm, and the outer delimiting membrane becomes the investing membrane of the ascospore (pls. 19; 20; 21, A; 22, A, B). The ascospore becomes elliptical in longitudinal sections as soon as the formation of the endospore is initiated (pl. 2, E; upper spore of plate 19. The image of the middle spore of plate 19 is essentially a cross section). The endospore is at first thickest in the median region of the spore (pls. 19, upper spore; 21, A). The endospore increases in thickness as the ascus matures until it reaches a uniform width of approximately 1 to 1.5μ (pls. 22, A, B; 23). The endospore remains electron-transparent until the ascospore is nearly mature. Then the inner and outer regions increase in electron-density (pl. 24, B). Since the ascospore illustrated in plate 24, B, represents the most mature stage that could be satisfactorily fixed and sectioned, further changes in the appearance of the endospore could not be followed.

The mucilaginous perisporal sac is formed between the investing membrane and the endospore approximately at the stage at which the endospore reaches its maximum width (pls. 22, B; 23). In the initial stages of formation, the perisporal sac forms a halo around the endospore that is thicker in the median part of the spore than at the poles (pl. 2, F). In some cases, the perisporal sacs of adjacent ascospores are initially confluent (pl. 25, A), but in the terminal stages of ascospore development the perisporal sacs are distinct (pl. 24, A). In the thin sections, the perisporal sac becomes somewhat irregular in outline during development (pl. 24, A). The characteristic unilateral orientation of the perisporal sac surrounding the mature ascospore (van Brummelen, 1967) can be seen with light (pl. 3, A, B, C) and electron microscopy (pl. 24, B).

Pigment granules are deposited within the perisporal sac in the terminal stages of ontogeny of the spore. The pigment granules are electron-dense structures (pls. 24, B; 25, A, B) comparable to those described by Delay (1966) in the wild-type strain of *Ascobolus immersus*. In the more mature spores of $KMnO_4$-fixed apothecia, the granular nature of the pigment granules is not always preserved (pl. 26, A).

The concentration of pigment granules is greatest adjacent to the epispore (pls. 3, B, C; 24, B; 25, A, B), but isolated clusters of pigment granules do occur in the outer regions of the perisporal sac (pls. 3, B; 24, B; 25, B). Evidently, the longitudinal striations on the surface of the ascospore of *A. stercorarius* (van Brummelen, 1967) are due to the deposition of the pigment granules in a pattern characteristic for the species. This could not be verified, however, because of the resistance of the mature ascospores to fixation, embedding, and thin sectioning.

Numerous lipid globules are formed on the surface of the perisporal sac during the latter stages of ontogeny, (pls. 3, D; 24, A, B; 25, A; 26, B). The lipid globules can be detected with interference-contrast optics (pls. 2, F; 3, A), and positive tests were obtained with Sudan black B (pl. 3, D) and with the OsO_4 technique (Jensen, 1962). The lipid globules are concentrated on the surface of the perisporal sac outside the investing membrane, and are usually enclosed by the tonoplast of the vacuole of the epiplasm (pl. 26, B). However, because of the highly contorted outlines of the tonoplast of the epiplasmic vacuole and of the investing membrane, the exact location of the lipid globules cannot always be determined (pl. 22, B). The lipid globules often project into the vacuole of the epiplasm but are surrounded by the tonoplast (pl. 25, A). At maturity the lipid globules disappear from the margin of the perisporal sac.

The epispore in *A. stercorarius* forms between the endospore and the perisporal sac approximately at the stage at which the deposition of pigment granules is initiated within the perisporal sac (pls. 24, A; 25, A). Cross sections of the epispore at an early stage of development indicate a spinose surface structure (pl. 25, A, right spore), but oblique to tangential sections indicate that the epispore is porous on the outer surface and essentially solid adjacent to the endospore (pl. 25, A).

The electron-transparent zone between the epispore and the perisporal sac is not evident at this porous stage (pl. 25, A).

Early in this study, the porous structure of the earlier stages of development of the epispore was considered to be an artifact; however, since this configuration has been seen in apothecia fixed by a variety of procedures using both $KMnO_4$ and glutaraldehyde and OsO_4, the porous structure is now considered to be real.

The porous layer increases in electron-density and becomes striated (pl. 25, B), and an inner, solid, electron-dense layer is deposited within the striated layer adjacent to the endospore (pl. 25, B). A layer of greater electron-density separates the outer striated zone and the inner solid layer (pl. 25, B). The outer striated zone consists of three electron-transparent bands and three alternating, somewhat discontinuous, electron-dense bands (pl. 25, B). At this stage, the electron-transparent layer in the perisporal sac adjacent to the epispore is obvious, especially in those regions containing the pigment granules (pl. 25, B).

Favorable thin sections of the epispore of the most mature ascospores that could be fixed and sectioned indicate that the broken electron-dense bands of the outer layer of the immature epispore become continuous, and that the inner, solid, electron-dense layer becomes striated (pl. 26, A). The most mature epispore observed consisted of seven electron-dense bands alternating with six electron-transparent bands (pl. 26, A). The outer, median, and inner electron-dense bands are broader and more electron-dense than are the other four electron-dense bands (pl. 26, A).

As noted by Delay (1966) in *A. immersus,* there is an electron-transparent region (pls. 25, A [left spore], B; 26, A) adjacent to the outer electron-dense band of the epispore. This layer becomes evident at the time the pigment granules are deposited in the perisporal sac (compare the left and right spores in pl. 25, A) and is present in the most mature spores that can be sectioned (pl. 26, A). Whether it is properly classified as a part of the perisporal sac or as a part of the epispore was not determined. However, since it has the same granular fine structure as the perisporal sac (pl. 26, A), I consider it is probably the inner layer of the perisporal sac.

The mature perisporal sac is easily dislodged from the epispore and endospore portions of the ascospore wall (pl. 3, F) and does not seem to be an integral part of the ascospore.

There is a reduction in the amount of glycogen within the sporoplasm soon after it is completely separated from the epiplasm (pls. 17, 18). Concurrent with the deposition of the endospore, glycogen vanishes from the electron micrographs of the sporoplasm (pls. 19, 20) and cannot be detected by the histochemical procedures utilized. In early periods of endospore formation, short, numerous fragments of the endoplasmic reticulum occur in the vicinity of the nucleus (pls. 19, 20, 21). In reality these short fragments probably represent a loose, interconnected, endoplasmic reticulum. In subsequent stages, the endoplasmic reticulum within the sporoplasm becomes less branched (pls. 22, A, B; 23; 24, A). In the thin sections of the nearly mature ascospores, cisternae of the endoplasmic reticulum are not evident (pl. 24, B); however, this is possibly due to the poor fixation of the sporoplasm at this stage.

At ascospore delimitation, the mitochondria in thin sections of the sporoplasm are spherical and tubular to dumbell-shaped (pls. 14, 16). As the ascospore matures, the tubular and spherical mitochondria become predominant (pls. 17; 18; 19; 20; 21, A, B; 22, A, B; 23; 24). Vacuoles are rare in the sporoplasm and, if present, occur only during the immature stages (pl. 21, A). Structures tentatively identified as autophagic vacuoles are present in the sporoplasm of spores with well-defined endospore and subsequent stages (pl. 22, B).

Lipid globules can be detected in maturing ascospores in apothecia treated with the OsO_4 or Sudan black B (pl. 3, D) techniques; however, lipid globules cannot be identified in the existing electron micrographs. Possibly the electron-dense bodies in the sporoplasm of the ascospore illustrated in plate 24, B, are lipid globules, but they are not typical of lipid globules detected in other structures (e.g., in the paraphysis in pl. 13). Rather, the electron-dense bodies within the sporoplasm of the ascospore of plate 24, B, more closely resemble autophagic vacuoles than lipid globules.

Two conspicuous structures occur consistently in the sporoplasm of the ascospores. These structures appear first in ascospores with developing endospore layers (pl. 27, A, B) and persist during all subsequent stages of ontogeny that were satisfactorily fixed and sectioned. One such structure is usually spherical and has an electron-dense, fine-grained interior surrounded by electron-dense granules (pl. 27, B). The other is composed of parallel fibers that often seem paired (pl. 27, A, B). The fibrous structure is often elongated in the electron micrographs. Both structures are preserved only in glutaraldehyde-OsO_4-fixed apothecia (pl. 27, A, B). In the sporoplasm of ascospores fixed in $KMnO_4$, regions devoid of organelles and membranes, and slightly more electron-transparent than the remaining cytoplasmic ground substance of the sporoplasm, are very probably the sites of the fibrous or fine-granular structures in the living spores (arrows in pls. 19; 20; 21, A). Spherical structures were noted by light microscopy in the sporoplasm of walled ascospores stained with acetocarmine. The similarity in form indicates that these may be granular structures noted by electron microscopy. The chemical composition of the granular and fibrous structures was not determined, but they were not differentiated by any of the techniques utilized to distinguish lipids or glycogen.

The fibrous structures demonstrated in the ascospores of our isolate of *A. stercorarius* appear identical to the fibrous structures described by Zachariah (1970) in the hyphal elements of the excipulum of the apothecia of a different isolate of *A. stercorarius*. Zachariah (1970) stated, however, that filament bundles were not observed in the sporoplasm. Beck, Decker, and Greenawalt (1970) observed "striated inclusions" resembling the fibrous structures observed by Zachariah (1970) and in the present study. Beck et al. (1970) noted that the striated inclusions occurred in both the cytoplasm and nucleoplasm of the somatic hyphae arising from the conidia of *Neurospora crass* Shear & Dodge. Such structures were observed when the conidia were germinated on media containing sucrose or glucose but not on media containing mannitol or lacking a carbohydrate supplement. Beck et al. (1970) noted that similar inclusions have been reported from a variety of organisms and organelles but that little information was available as to the composition and function of such structures.

During the latter stages of ontogeny, the ascospores are arranged in an uneven single row (pls. 2, E, F; 3, A). However, in the terminal stages of ascus enlargement, immediately prior to ascospore discharge, there is sufficient space for a partial double row of ascospores in many asci (pl. 3, C).

Epiplasm

The epiplasmic vacuoles are initially most abundant between the ascospores and ascus wall (pls. 18, 19, 20). As the endospore layer of the ascospore increases in thickness, the vacuoles increase in number (pls. 18; 19; 20; 21, A). The epiplasmic vacuoles become more abundant adjacent to the investing membranes of the ascospores (pl. 21, A) and then enlarge and coalesce in the interascospore zone (pls. 21, B; 22, A) and extend around the ascospore (pl. 22, A). The eventual result is the differentiation of the epiplasm into a large central vacuole enclosing the eight developing ascospores and a narrow layer of cytoplasm along the periphery of the sporogenous zone (pl. 23), which is somewhat broader at the apex (pls. 2, F; 21, B).

In the initial stages of vacuolation of the sporogenous zone, there is a small amount of cytoplasm between the investing membrane of the spore and the tonoplast (pl. 22, A). However, except when separated by lipid globules, the tonoplast and investing membrane become closely and evenly aligned when the perisporal sac is initiated, to form a contorted, double membrane system (pl. 22, B). In all favorable thin sections of subsequent stages, this close association is visible (pl. 26, B).

In material fixed in $KMnO_4$, the epiplasmic vacuoles have an electron-dense, granular content during vacuolation (pls. 21, B; 22, A; 23). As the perisporal sac develops, the granular material of the vacuole concentrates in the vicinity of the ascospores and the ascus wall (pl. 24, A) and eventually disappears (pl. 25, B). In asci fixed in glutaraldehyde-OsO_4, the electron-dense, granular content of the vacuoles is not evident at any stage (pls. 22, B; 24, B).

The ultrastructural features of the epiplasm at the time of its differentiation from the sporoplasm of the ascospores are identical with those described for the ascus cytoplasm of the sporogenous zone. During vacuolation of the epiplasm, there is an apparent increase in glycogen granules (pls. 13; 15; 21, B; 22, B; 24, A, B). The Bauer-Feulgen technique (Glick, 1949) indicates that the peripheral cytoplasm of the epiplasm after vacuolation has the highest concentration of glycogen. These results are in conformity with observations made with the electron microscope (pls. 21, B; 22, B; 24, A, B). In the terminal stages of ascus maturation, glycogen, as indicated by the intensity of the Bauer-Feulgen test, disappears from the epiplasm. Such stages of ascus development could not be studied satisfactorily by electron microscopy but are in agreement with other studies of the species of the genus *Ascobolus* (van Brummelen, 1967). Although lipid globules are rare in electron micrographs of the epiplasm of the distal portion of the ascus, scattered, positive-staining globules were detected in materials stained with Sudan black B. Lipid globules are relatively abundant in such material in the basal portion of the ascus in the terminal stages of maturation.

Within the epiplasm, spherical and cylindrical mitochondria become increasingly more common as the ascus matures (pls. 19; 20; 21, A; 23); however, the characteristic dumbbell-shaped mitochondria of earlier stages do persist for some time (pls. 19; 21, A). Short segments of the endoplasmic reticulum are especially abundant during formation of the endospore. During deposition of the endospore there is a consistent parallel association of segments of endoplasmic reticulum with the investing membrane (pls. 20; 21, A). A similar spatial relation occasionally occurs at earlier stages (pl. 18), but the perpendicular junction of the endoplasmic reticulum and the investing membranes described by Delay (1966) in *A. immersus* was not observed in *A. stercorarius.*

Ascus Wall

As noted by Delay (1966) in *A. immersus,* the lateral and apical regions of the wall of vacuolated diploid asci of *A. stercorarius* and of asci at subsequent stages consist of a thin, outer, electron-dense layer and a broad, inner, electron-transparent layer of uniform thickness (pls. 5, B; 7, A, B; 9; 10, A, B; 12, A, B, C; 13; 15; 16; 17; 18; 19; 20; 21, A, B; 22, A; 23; 24, A, B; 27, A; 28, A, B; 29, A, B, C). The outer, electron-dense layer is not evident in the walls of young, diploid asci (pl. 5, A), dikaryotic asci (pl. 4, B), or croziers (pl. 4, A).

The average thickness of the lateral ascus wall adjacent to the sporogenous zone in three asci at the four-nucleate stage (e.g., pl. 10, A, B), was 180 mμ, as measured from electron micrographs of $KMnO_4$-fixed apothecia. The ascus wall peripheral to the sporogenous zone in three asci containing ascospores with developing epispore (e.g., pl. 24, A) was 200 mμ, as measured from electron micrographs of $KMnO_4$-fixed apothecia. Thus there is no significant increase in thickness of the lateral ascus wall during the latter stages of ontogeny.

Differentiation of the operculum is initiated at the apex during the development of the endospore (pl. 21, B). The first indication of the operculum is the appearance of an annular indentation in the inner surface of the wall at the margin of the operculum (pl. 21, B). The indentation is approximately 0.6 to 1.0 μ in width. The wall in the indentation is about 150 to 180 mμ in thickness at the inward edge and approximately 45 to 65 mμ in thickness at the outward edge. Adjacent to the annular indentation and toward the center of the operculum, an annular thickening of the inner wall develops as the ascus matures (pls. 21, B; 25, C). The thickened region becomes approximately 350 mμ in width, as measured from the electron micrographs of the most mature asci that could be fixed and sectioned adequately. The median portion of the operculum, located at the apex of the ascus, also increases in thickness, becoming approximately 225 to 275 mμ in the more mature asci. Like the lateral ascus wall, the operculum is composed of an inner, thick, electron-transparent layer and an outer, thin, electron-dense layer (pls. 21, B; 25, C).

In the early stages of development, the operculum is evenly crescent-shaped (pl. 21, B), but it becomes flattened to subumbonate in the later stages of ontogeny (pls. 3, B, E; 25, C). When viewed by light microscopy in water mounts, the margins of the opercula of mature asci are often depressed prior to ascospore discharge (pl. 3, B). This form of the operculum is maintained after ascospore release (pl. 3, E).

At the time of discharge, the separation of the operculum and ascus wall takes place, apparently at the outer edge of the annular indentation. The site of separation could not be detected by electron microscopy but is indicated by the shape of the operculum following discharge (pl. 3, E) and by the fact that this is the thinnest portion of the operculum (pls. 21, B; 25, C). As noted by van Brummelen (1967), the operculum often remains attached to the ascus wall at one point (pl. 3, E), but at times it is completely detached at discharge.

DISCUSSION

The synthesis of glycogen granules in the ascus of *A. stercorarius* seems to be associated with a concentric membrane complex that is preserved by glutaraldehyde and OsO_4. Glycogen granules have been found associated with smooth membranes in several types of animal cells. As examples, Flaks (1968) reported that glycogen occurred in the vicinity of concentric, smooth-surfaced membranes in rat hepatomas; and glycogen membrane complexes composed of pairs of parallel, smooth membranes and mono-particulate glycogen have been described in mouse muscle tissue by Garant (1968).

Calonge (1969) has reported the presence of glycogen membrane complexes in the somatic hyphae of *Schlerotinia fructigena* Aderh. & Ruhl. He found that glycogen granules were associated with smooth membrane complexes that were not preserved in $KMnO_4$. The membrane complexes described by Calonge (1969) differed morphologically from those found in *A. stercorarius*. In *Sclerotinia fructigena,* the spherical membranes were scattered within vacuoles rather than arranged in concentric pairs as in *A. stercorarius*.

The frequent association of glycogen granules with the concentric membrane complexes in the asci of *A. stercorarius,* especially in the diploid and four- and eight-nucleate stages, indicates that the concentric membrane complexes may be involved in the synthesis of glycogen granules. This postulate is supported by the observations of Calonge (1969) and by the reports of glycogen membrane complexes in animal tissues (e.g., Flaks, 1968; Garant, 1968).

It has been proposed that the formation and enlargement of vacuoles is the mechanism controlling the movement of cytoplasm within hyphae (Buller, 1933), sporangia and sporangiophores (Buller, 1933), croziers (Hung and Wells, 1971), homobasidia (Buller, 1933; Corner, 1948; Wells, 1965), and heterobasidia (Wells, 1964). It seems generally accepted that the turgor pressure developing within the vacuole of the mature epiplasm is responsible for the forcible discharge of ascospores in many Ascomycetes (Ingold, 1966). Previous investigators of ascus development have reported the development of basal and apical vacuoles (de Bary, 1863, 1866; Harper, 1899). Fraser and Brooks (1909) noted the formation of internuclear vacuoles within the sporogenous zone in eight-nucleate asci of *Ascobolus furfuraceus*. Prior to ascospore delimitation, such internuclear vacuoles seemed to function, according to Fraser and Brooks (1909), to separate the cytoplasm surrounding the individual nuclei.

In *A. stercorarius* during ascus ontogeny, the increases and decreases in the number and dimension of vacuoles within the ascus protoplast follow an orderly pattern that can be observed with both light and electron microscopy. As stressed by Robertson (1968), it is impossible, on the basis of available information, to determine whether vacuolation is a cause or a consequence of the movement of the cytoplasm. However, the vacuolar pattern during ascus ontogeny in *A. stercorarius* is consistent with the concept that vacuoles function to concentrate the cytoplasm in that portion of the ascus in which growth, nuclear divisions, and ascosporogenesis take place.

The elongation of the dikaryotic ascus and the young diploid ascus is accompanied by the enlargement of vacuoles in the basal region. Prior to division I, a cluster of small apical vacuoles arises in the distal portion of the ascus between the cytoplasm adjacent to the diploid nucleus and the apical cytoplasm. It is at this stage that the septa at the base of the ascus become occluded with electron-dense plugs. These would apparently restrict the movement of additional cytoplasm into the ascus from the adjacent segments of the ascogenous hypha. Obviously, subsequent development of the ascus must take place without a substantial increase in the amount of cytoplasm within the ascus. During divisions I to III, the pattern of the distribution of vacuoles established in the diploid ascus is maintained, except that internuclear vacuoles often appear in the sporogenous zone at the eight-nucleate stage. Increase in size of the ascus is accompanied by increased vacuolation of the ascus protoplast, especially in the basal portion. Following delimitation and reorientation of the ascospores, the apical vacuoles disappear, and progressive vacuolation of the epiplasm is initiated. Vacuoles apparently do not occur in mature ascospores. Following deposition of the endospore, the ascus protoplast external to the investing membranes of the spores becomes completely vacuolated except for a thin, dense layer of cytoplasm adjoining the ascus wall. This phase of vacuolation is accompanied by a rapid increase in size of the ascus, both length and width increasing twofold. Presumably, the turgor pressure responsible for the forcible discharge of the ascospores develops within the large vacuole of the mature epiplasm.

There is general agreement among the authors of the more recent reports of ascus ontogeny by electron miscroscopy, that the ascospore is differentiated by double ascospore-delimiting membranes and that the ascospore wall is formed between the two delimiting membranes (Bandoni et al., 1967; Beckett and Crawford, 1970; Beckett et al., 1968; Bracker, 1967, 1969; Bracker and Williams, 1966; Carroll, 1967*a,* 1969; Conti and Naylor, 1960; Delay, 1966; Furtado and Olive, 1970; Greenhalgh and Evans, 1968; Greenhalgh and Griffiths, 1970; Marquardt, 1963; Oso, 1969; Reeves, 1967; Schrantz, 1966, 1970*b;* Wilsenach and Kessel, 1965; Zickler, 1970). The results of the present study of *A. stercorarius* are in accordance with this concept.

Oso (1969) has reported that the double ascospore-delimiting membranes appeared first in the distal portion of the ascus and developed basally to form a sac around the sporogenous zone in asci of *A. crenulatus.* In contrast, several other studies of ascus ontogeny have indicated that the ascospore-delimiting membranes appeared first along the lateral ascus wall (Bracker, 1969; Carroll, 1967*a,* 1969; Reeves, 1967) and finally above and below the sporogenous zone (Carroll, 1967*a,* 1969; Reeves, 1967). The latter sequence was observed in the present study and is, in my opinion, more probable, since several of the electron micrographs used as illustrations by Oso (1969) seem to be oblique sections.

Both Carroll (1967*a,* 1969) and Reeves (1967) noted that the double ascospore-delimiting membranes are not continuous. The general concept proposed by Carroll (1967*a,* 1969) and Reeves (1967) conforms to that gained in the present study; however, I believe that the process of delimitation involves inward growth of the delimiting membranes, possibly accompanied by some invagination of the membranes. In my opinion, such a concept is necessary to account for the similarities in ultrastructural characteristics of the sporoplasm and epiplasm immediately after delimitation. In *A. stercorarius,* the membranes and organelles seem to have the same distribution in the sporoplasm and epiplasm. In *A. immersus,* Delay (1966) noted that the cytoplasmic structures were even less densely distributed in the sporoplasm than in the surrounding epiplasm. Also supporting the concept of inward growth of the double ascospore-delimiting membranes are the observations made during the present study and those made by Carroll (1969) that nuclear blebbing persists during ascospore delimitation and can be consistently detected, at least in *A. stercorarius,* until the initiation of the endospore between the delimiting membranes.

Several postulates have been proposed to account for the origin of the double ascospore-delimiting membranes. Reeves (1967) and Wilsenach and Kessel (1965) suggested that the delimiting membranes are derived from the endoplasmic reticulum. Bandoni et al. (1967), Bracker (1969), and Greenhalgh and Griffiths (1970) presented evidence that the delimiting membranes originate from the plasma membrane of the ascus protoplast. Such a hypothesis is supported by the similarities in structure of the delimiting membranes and the ascus plasma membrane noted in the present study, and also noted by Bracker (1967, 1969), Delay (1966), Furtado and Olive (1970), Greenhalgh and Griffiths (1970), and Schrantz (1970*b*), and by the continuities between the delimiting membranes and the plasma membrane clearly demonstrated by Bracker (1969) and by Greenhalgh and Griffiths (1970). Continuities between the two membrane systems could not, however, be demonstrated in the present study of *A. stercorarius.*

The evidence obtained in the present study is in accordance with the nuclear bleb hypothesis proposed by Carroll (1967*a,* 1969) and supported by Beckett and Crawford (1970). It should be possible, if Carroll's concept is correct, to demonstrate a gradual transformation of the associated segment of endoplasmic reticulum and nuclear envelope, each composed of two membranes, into two, distinctly tripartite, asymmetrical membranes. In Carroll's (1967*a*) report and in the present study there are indications of membrane modification during nuclear blebbing; however, conclusive evidence of the required transformation is not available.

A third hypothesis, recently proposed by Bracker (1969), deserves careful consideration. Bracker suggested that Golgi apparatus-like cisternae are transformed from endoplasmic reticulum-like to plasma membrane-like images forming vesicles that contribute to the growth of the delimiting membranes. Such a hypothesis is in accordance with the concept developed by Grove et al. (1970) and Grove and Bracker (1970) of membrane transformations that occur during hyphal tip growth.

In my opinion, the mechanism of formation of the double ascospore-delimiting membranes is in need of additional study, and definite conclusions should not be reached on the basis of available information. Although the present study supports Carroll's (1967*a*, 1969) proposal, the evidence is not conclusive. My efforts to obtain better fixation of the membranes within the asci of *A. stercorarius* were not successful, perhaps because the asci at the appropriate stages of ontogeny are completely embedded within a gelatinous matrix. The illustrations of continuities between the plasma membrane and the delimiting membranes presented by Bracker (1969) and by Greenhalgh and Griffiths (1970) are explicit, and Bracker's (1969) proposal that the delimiting membranes are derived from vesicles originating from Golgi apparatus-like cristernae is most appealing.

As would be expected, there are a number of similarities between ascospore development in *A. stercorarius* and in *A. immersus,* as described by Delay (1966). In both species the electron-transparent endospore is formed prior to the deposition of the perisporal sac. Subsequently, the laminated epispore is formed between the endospore and the investing membrane (Delay, 1966). Seemingly, the mucilaginous perisporal sac and epispore mature concurrently in both *A. immersus* (Delay, 1966) and *A. stercorarius.* In contrast, Oso (1969) reported that epispore formation in *A. viridulus* followed deposition of the electron-transparent endospore, and that the mucilaginous layer (=perisporal sac) was formed outside the investing membrane. The same sequence of development of the ascospore wall and perisporal sac as reported in the present study, and by Delay (1966), was noted by Reeves (1967) in *Pyronema domesticum.* However, other studies (Bandoni et al., 1967; Beckett et al., 1968; Greenhalgh and Evans, 1968; Schrantz, 1907*b*) of taxonomically diverse species indicate that the sequence of ascospore wall formation within the Ascomycetes is possibly not uniform.

A number of students of ascospore ontogeny have commented on the roles of the epiplasm and sporoplasm in the deposition of the perisporal sac and of the various layers of the ascospore wall (Chadefaud, 1960; Delay, 1966; Marquardt, 1963; Moore, 1963). In the present study of *A. stercorarius* no evidence was noted that would permit a determination of the source of the components of the endospore, epispore, or perisporal sac. Delay (1966) presented some evidence that suggested that the mucilaginous substance of the perisporal sac was derived from the epiplasm. However, she concluded that it was not possible to determine from the static images of the electron micrographs whether the substances forming the endospore and epispore originated in the epiplasm or in the sporoplasm. Although I am in agreement with Delay's conclusion, it is profitable, I believe, to compare the sequences of deposition of wall layers in the ascospores of *A. immersus* (Delay, 1966), *A. stercorarius,* and some other species (Beckett et al., 1968; Marquardt, 1963; Moore, 1963; Reeves, 1967) with the sequence in other fungal structures. For example, wall formation of the somatic cells of *Sporobolomyces roseus* Kl. &

van Niel (Prusso and Wells, 1967) and of the sporangiospores of *Gilbertella persicaria* (Eddy) Hesseltine (Bracker, 1968) is centripetal. Centripetal wall formation seems to be the rule in the fungi; the only exceptions apparently occur during ascospore wall formation of some species. Conceivably, this unique sequence of ascospore wall formation is indicative of a role of the epiplasm in wall deposition.

The epispore of *A. immersus,* as illustrated by Delay (1966), consists of at least 15 electron-dense bands. The third electron-dense band external from the plasma membrane of the sporoplasm is broader than the remaining laminae. Oso (1969) published electron micrographs illustrating a portion of a mature ascospore of *A. viridulus* in which the epispore apparently consists of five electron-dense layers of uniform thickness. Thus, the epispores of the three species of *Ascobolus* examined to date are laminated, but the number and dimensions of the electron-dense laminae are distinct for each species.

In Delay's (1966) study of a wild-type strain with pigmented ascospores and of a mutant strain of *A. immersus* with colorless spores, she found that electron-dense granules were deposited in the perisporal sac of the mature ascospores of the wild-type strain. Since these granules were lacking in the colorless ascospores of the mutant strain, it is probable that they are responsible for the color of the wild-type ascospores. In *A. stercorarius* granules indentical to those demonstrated by Delay (1966) in the wild-type strain of *A. immersus,* appear within the perisporal sac of maturing ascospores. In addition, ascospores from which the perisporal sac has been removed are colorless when examined by light microscopy. These findings support Chadefaud's (1942) belief that the pigment of the ascospores of the species of *Ascobolus* is deposited not in the spore wall proper but in the perisporal sac.

In *A. stercorarius* segments of the endoplasmic reticulum were oriented parallel to the investing membranes of immature ascospores. A similar orientation was noted by Delay (1966) in *A. immersus.* Apparently this association occurs at approximately the same stage of ascospore ontogeny in both species, i.e., during the deposition of the endospore. Delay (1966) also noted at a subsequent stage of ontogeny, a perpendicular orientation of segments of smooth endoplasmic reticulum that were continuous with the investing membrane. Such membrane configurations were not seen in the electron micrographs of *A. stercorarius.* Reeves (1967) also reported that elements of the endoplasmic reticulum were aligned parallel to the developing endospore in *Pyronema domesticum;* however, the segments of endoplasmic reticulum were within the sporoplasm, not in the epiplasm as in *A. immersus* and *A. stercorarius.* Reeves (1967) suggested that the association of the endoplasmic reticulum with the endospore in *P. domesticum* might indicate that the endoplasmic reticulum functioned in wall formation. A similar suggestion was made by Lynn and Magee (1970) in their study of *Saccharomyces cerevisae.*

Lomasomes adjacent to the plasma membrane of the sporoplasm or to the investing membrane were observed infrequently in the electron micrographs of *A. stercorarius.* Neither Delay (1966) nor Oso (1969) emphasized the presence of lomasomes during the formation of the ascospore wall in the species of *Ascobolus* studied by them. In my opinion, lomasomes were not present in a sufficient number of thin sections of developing ascospores to justify postulating that lomasomes play a role in the formation of the walls of ascospores of *A. stercorarius,* as suggested by Wilsenach and Kessel (1965) in their study of *Talaromyces vermiculatus* and Carroll (1969) in his study of *Saccobolus glaber.*

The results of recent electron microscopic studies of mitosis (Girbardt, 1968; Motta, 1969) and meiosis (Lu, 1967; Lerbs and Thielke, 1969; Thielke, 1968, 1969) in the Basidiomycetes and of mitotic and meiotic divisions in the Ascomycetes (Beckett and Crawford, 1970; Robinow and Caten, 1969; Robinow and Marak, 1966; Schrantz, 1967, 1970*a;* Wells, 1970; Zickler, 1970) indicate significant differences in the behavior of the nuclear envelope during nuclear divisions in the two classes. These studies also reveal that the polar, centriole-like structures present during nuclear divisions are unique in each class. Although the perfect reproductive structures of the two classes possess some characteristics in common, the mechanisms of ascospore and basidiospore formation are quite dissimilar. The free cell formation of the ascospores seems to be a unique process, and is related to basidiospore formation only to the extent that the resulting spores are both perfect spores.

Although the studies of *Endomycopsis platypodis* Baker & Kreger-van Rij and *Endomycopsis monospora* Saito (Kreger-van Rij and Veenhuis, 1969) and of *Puccinia graminis* Pers. and *Puccinia recondita* Rob. ex Desm. (Ehrlich, Ehrlich, and Schafer, 1968) indicate that there may be intermediate types, the typical septa in the somatic thalli of the Ascomycetes and Basidiomycetes are distinct morphologically (Bracker, 1967) and, in part, functionally. In addition to a structural function, the basidiomycete septal spore apparatus seems to prevent nuclear migration betwen hyphal segments (Giesy and Day, 1965).

The differences in septal structure and apparent function, mechanics of nuclear divisions, and perfect reproductive structures indicate, in my opinion, that the phylogenetic affinities between the Ascomycetes and Basidiomycetes are not as close as many of our present taxonomic systems indicate. Although the two classes have a number of characteristics in common, such as a dikaryotic nuclear state during some stages of the life cycle, recent studies utilizing the electron microscope indicate that they are more divergent taxa than is generally believed.

SUMMARY

Except for the absence of Woronin bodies during all stages of ontogeny, and for the presence of electron-dense masses adjacent to the nuclei in dikaryotic and diploid asci, the ascus protoplast of *Ascobolus stercorarius* has essentially the same ultrastructural features as the somatic hyphae of most other species of Ascomycetes. Vacuoles are present in the basal portion of the penultimate cell of the crozier and the dikaryotic ascus. Following karyogamy, basal vacuoles increase in number and dimension as clusters of relatively small vacuoles are formed in the distal portion of the ascus from smooth, netted, endoplasmic reticulum. This distribution of vacuoles within the ascus protoplast persists until ascospore delimitation. At delimitation the apical vacuoles disappear, and the epiplasm becomes increasingly vacuolated until only a relatively thin, peripheral layer of cytoplasm remains adjacent to the ascus wall.

In the diploid ascus, an apical zone devoid of most membranes and organelles is evident. Apical vesicles, apparently associated with apical growth of the ascus, are present in this zone, which persists until after ascospore delimitation. Most of the mitochondria have a dumbbell-like morphology in thin sections of the older diploid asci, rather than the long and sinuous form of earlier stages. The two septa at the base of the ascus become plugged in the diploid nuclear stage.

Glycogen appears initially in the ascus protoplast during the diploid stage. It increases in amount in the ascus cytoplasm prior to ascospore delimitation and in the epiplasm following delimitation in all asci that can be studied with the electron microscope. Histochemical tests indicate that glycogen disappears in the epiplasm prior to ascospore discharge. The synthesis of glycogen seems to be associated with concentric membrane complexes, which are composed of asymmetrical, tripartite membranes. These concentric membrane complexes are apparently derived from the plasma membrane of the ascus protoplast.

The double ascospore-delimiting membranes that are composed of two asymmetrical, tripartite membranes develop laterally along the ascus wall during the four-nucleate stage. The evidence obtained supports the hypothesis that the delimiting membranes are formed from nuclear blebs; however, distinct electron micrographs illustrating the membrane transformations required for this hypothesis were not obtained. Following division III, the discontinuous delimiting membranes developed above and below the sporogenous portion of the ascus.

The ascospores are delimited by the inward growth, possibly accompanied by some invagination, of the double ascospore-delimiting membranes. The delimiting membranes separate into eight portions that surround the eight nuclei. No evidence was obtained indicating that the astral rays function in ascospore delimitation.

After the rearrangement of the ascospores, the endospore is deposited between the delimiting membranes of the ascospores. The ascospores then become elliptical. The inner ascospore-delimiting membrane becomes the plasma membrane of the ascospore, and the outer delimiting membrane becomes the investing membrane of the spore. The mucilaginous perisporal sac is formed between the endospore and the investing membrane. Lipid globules are especially abundant adjacent to the outside of the investing membrane during the latter stages of ascospore ontogeny, but disappear when the ascospore matures. During growth of the perisporal sac, the laminated epispore develops between the perisporal sac and the endospore. Pigment granules are deposited within the perisporal sac and become especially abundant adjacent to the epispore. At maturity, only scattered clusters of pigment granules remain in the outer region of the unilateral perisporal sac. The sporoplasm has essentially the same ultrastructural characteristics as the cytoplasm of diploid asci, except that spherical to tubular mitochondria become predominant in the latter stages of ontogeny. Two unidentified structures are consistently present in the sporoplasm of the more mature ascospores.

The wall of diploid asci with apical vacuoles, and asci at all subsequent stages, consists of a thin, outer, electron-dense layer and a broad, inner, electron-transparent layer. The margin of the operculum is defined by an annular indentation of the inner wall. An annular thickening develops in the inner wall adjacent to the indentation. The operculum becomes flattened to subumbonate at maturity; and separation of the operculum at the time of ascospore discharge occurs, apparently at the outer edge of the annular indentation.

LITERATURE CITED

ANDRUS, C. F., and L. L. HARTER

1933. Morphology of reproduction in *Ceratostomella fimbriata*. J. Agr. Res. 46:1059–1078.

1937. Organization of the unwalled ascus in two species of *Ceratostomella*. J. Agr. Res. 54: 19–46.

BANDONI, R. J., A. A. BISALPUTRA, and T. BISALPUTRA

1967. Ascospore development in *Hansenula anomala*. Can. J. Bot. 45:361–366.

BARY, A. DE

1863. Über die Fruchtentwicklung der Ascomyceten. Wilhelm Engelmann, Leipzig. 38 p.

1866. Morphologie und Physiologie der Pilze, Flechten und Myxomyceten. Wilhelm Engelmann, Leipzig. 316 p.

BECK, D. P., G. L. DECKER, and J. W. GREENAWALT

1970. Ultrastructure of striated inclusions in *Neurospora*. J. Ultrastruct. Res. 33:245–251.

BECKETT, A., R. BARTON, and I. M. WILSON

1968. Fine structure of the wall and appendage formation in ascospores of *Podospora anserina*. J. gen. Microbiol. 53:89–94.

BECKETT, A., and R. M. CRAWFORD

1970. Nuclear behavior and ascospore delimitation in *Xylosphaera polymorpha*. J. gen. Microbiol. 63: 269–280.

BECKETT, A., and I. M. WILSON

1968. Ascus cytology of *Podospora anserina*. J. gen. Microbiol. 53:81–87.

BJÖRLING, K.

1941. Zur Kenntnis der Kernverhältnisse im Ascus von *Ascobolus stercorarius*. K. Fysiogr. Sällsk. Lund förh. 11:46–62.

BRACKER, C. E.

1967. Ultrastructure of fungi. Ann. Rev. Phytopath. 5:343–374.

1968. The ultrastructure and development of sporangia in *Gilbertella persicaria*. Mycologia 60:1016–1067.

1969. Sporogenesis in Mucorales and Ascomycetes. XI. Int. Bot. Congress, p. 20. (Abstr.) Seattle, Washington.

BRACKER, C. E., and C. M. WILLIAMS

1966. Comparative ultrastructure of developing sporangia and asci in fungi, p. 307–308. (Abstr.) *In* R. Uyeda [ed.] Sixth Int. Congr. Electron microscopy, Vol. II. Maruzen Co., Ltd., Tokyo.

BRENNER, D. M., and G. C. CARROLL

1968. Fine-structural correlates of growth in hyphae of *Ascodesmis sphaerospora*. J. Bacteriol. 95:658–671.

BRUMMELEN, J. VAN

1967. A world-monograph of the genera *Ascobolus* and *Saccobolus* (Ascomycetes, Pezizales). Persoonia Suppl. 1:1–260.

BULLER, A. H. R.

1915. Micheli and the discovery of reproduction in fungi. Trans. Roy. Soc. Can., Ser. III, Sec. IV, 9:1–25.

1933. The translocation of protoplasm through the septate mycelium of certain Pyrenomycetes, Discomycetes, and Hymenomycetes, p. 75–167. *In* A. H. R. Buller, Researches on fungi, Vol. 5. Longmans, Green and Co., London.

CALONGE, F. D.

1969. The occurrence of glycogen-membrane complexes in fungi: an electron microscope study. Protoplasma 67:79–85.

CARROLL, G. C.

1967*a*. The ultrastructure of ascospore delimitation of *Saccobolus kerverni*. J. Cell Biol. 33: 218–224.

1967*b*. The fine structure of the ascus septum in *Ascodesmis sphaerospora* and *Saccobolus kerverni*. Mycologia 59:527–532.

1969. A study of the fine structure of ascosporogenesis in *Saccobolus kerverni*. Arch. Mikrobiol. 66:321–339.

CHADEFAUD, M.

1942. Etudes d'asques, III: Le mécanisme de la pigmentation des ascospores chez *Ascobolus*. Bull. Soc. bot. Fr. 89:58–61.

1943. La formation des ascospores chez la Pézize inoperculée *Mniaecia jungermanniae* Boudier. Bull. Soc. bot. Fr. 90:207–209.

1960. Les végétaux non vasculaires, 1018 p. *In* M. Chadefaud and L. Emberger. Traité de Botanique. Vol. 1. Masson et Cie, Paris.

CLAUSSEN, P.

1907. Zur Kenntnis der Kernverhältnisse von *Pyronema confluens*. Ber. Deut. bot. Ges. 25: 586–590.

1912. Zur Entwicklungsgeschichte der Ascomyceten. *Pyronema confluens*. Z. Bot. 4:1–64.

CONTI, S. F., and H. B. NAYLOR

1960. Electron microscopy of ultrathin sections of *Schizosaccharomyces octosporus*. III. Ascosporogenesis, ascospore structure, and germination. J. Bacteriol. 79:331–340.

CORNER, E. J. H.

1929. Studies in the morphology of Discomycetes. II. The structure and development of the ascocarp. Trans. Brit. mycol. Soc. 14:275–291.

1948. Studies in the basidium. I. The ampoule effect, with a note on nomenclature. New Phytol. 47:22–51.

DANGEARD, P. A.

1894. La reproduction sexuelle des Ascomycètes. Botaniste 4:21–58.

DELAY, C.

1966. Etude de l'infrastructure de l'asque d'*Ascobolus immersus* Pers. pendant la maturation des spores. Ann. Sci. Natur. (Bot.) 7:361–420.

DENNIS, R. W. G.

1968. British Ascomycetes. J. Cramer, Stuttgart. 455 p.

EHRLICH, M. A., H. G. EHRLICH, and J. F. SCHAFER

1968. Septal pores in the Heterobasidiomycetidae, *Puccinia graminis* and *P. recondita*. Amer. J. Bot. 55:1020–1027.

EMERSON, R.

1958. Mycological organization. Mycologia 50:589–621.

FEDER, N., and T. P. O'BRIEN

1968. Plant microtechnique: some principles and new methods. Amer. J. Bot. 55:123–142.

FLAKS, B.

1968. Formation of membrane-glycogen arrays in rat hepatoma cells. J. Cell Biol. 36:410–414.

FRASER, H. C. I., and W. E. ST. JOHN BROOKS

1909. Further studies on the cytology of the ascus. Ann. Bot. 23:537–549.

FURTADO, J. S., and L. S. OLIVE

1970. Ultrastructure of ascospore development in *Sordaria fimicola*. J. Elisha Mitchell Sci. Soc. 86:131–138.

GAMUNDÍ, I. J., and M. E. RANALLI

1963. Apothecial development in *Ascobolus stercocarius*. Trans. Brit. mycol. Soc. 46:393–400.

GARANT, P. R.

1968. Glycogen-membrane complexes within mouse striated muscle cells. J. Cell Biol. 36:648–653.

GIESY, R. M., and P. R. DAY

1965. The septal pores of *Coprinus lagopus* in relation to nuclear migration. Amer. J. Bot. 52:287–293.

GIRBARDT, M.

1968. Ultrastructure and dynamics of the moving nucleus. Symp. Soc. exp. Biol. 22:249–260.

1969. Die Ultrastruktur der Apikalregion von Pilzhyphen. Protoplasma 67:413–441.

GJURASIN, S.
1893. Ueber die Kerntheilung in den Schläuchen von *Peziza vesiculosa* Bulliard. Ber. Deut. bot. Ges. 11:113–117.

GLICK, D.
1949. Techniques of histo- and cytochemistry. Interscience Publishers, Inc., New York, 531 p.

GREENHALGH, G. N., and L. V. EVANS
1968. The developing ascospore wall of *Hypoxylon fragiforme*. J. Roy. micros. Soc. 88:545–556.

GREENHALGH, G. N., and H. B. GRIFFITHS
1970. The ascus vesicle. Trans. Brit. mycol. Soc. 54:489–492.

GROVE, S. N., and C. E. BRACKER
1970. Protoplasma organization of hyphal tips among fungi: vesicles and Spitzenkörper. J. Bacteriol. 104:989–1009.

GROVE, S. N., C. E. BRACKER, and D. J. MORRÉ
1970. An ultrastructural basis for hyphal tip growth in *Pythium ultimum*. Amer. J. Bot. 57:245–266.

GWYNNE-VAUGHAN, H. C. I., and Q. E. BROADHEAD
1936. Contributions to the study of *Ceratostomella fimbriata*. Ann. Bot. 50:747–758.

HARPER, R. A.
1895. Beitrag zur Kenntniss der Kerntheilung und Sporenbildung im Ascus. Ber. Deut. bot. Ges. 13:(67)–(78).
1896. Ueber das Verhalten der Kerne bei der Fruchtentwickelung einiger Ascomyceten. Jahrb. wiss. Bot. 29:655–685.
1897. Kerntheilung und freie Zellbildung im Ascus. Jahrb. wiss. Bot. 30:249–284.
1899. Cell-division in sporangia and asci. Ann. Bot. 13:467–525.
1905. Sexual reproduction and the organization of the nucleus in certain mildews. Carnegie Inst. Wash. Publ. 37. 104 p.

HASHIMOTO, T., P. GERHARDT, S. F. CONTI, and H. B. NAYLOR
1960. Studies on the fine structure of microorganisms. V. Morphogenesis of nuclear and membrane structures during ascospore formation in yeast. J. biophys. biochem. Cytol. 7:307–310.

HUNG, C.-Y., and K. WELLS
1971. Light and electron microscopic studies of crozier development in *Pyronema domesticum*. J. gen. Microbiol. 66:15–27.

INGOLD, C. T.
1966. Aspects of spore liberation: violent discharge, p. 113–132. *In* M. F. Madelin [ed.] The fungus spore. Proc. 18th Symp. Colston Res. Soc., Butterworths, London.

JANCZEWSKI, E. V. G.
1871. Morphologische Untersuchungen über *Ascobolus furfuraceus*. Bot. Ztg. 29:257–262, 271–279.

JENSEN, W. A.
1962. Botanical histochemistry. W. H. Freeman and Co., San Francisco. 408 p.

KREGER-VAN RIJ, N. J. W., and M. VEENHUIS
1969. Septal pores in *Endomycopsis platypodis* and *Endomycopsis monospora*. J. gen. Microbiol. 57:91–96.

LERBS, V., and C. THIELKE
1969. Die Entstehung der Spindel während der Meiose von *Coprinus radiatus*. Arch. Mikrobiol. 68:95–98.

LU, B. C.
1967. Meiosis in *Coprinus lagopus:* a comparative study with light and electron microscopy. J. Cell Sci. 2:529–536.

LYNN, R. R., and P. T. MAGEE
1970. Development of the spore wall during ascospore formation in *Saccharomyces cerevisiae*. J. Cell Biol. 44: 688–692.

MARQUARDT, H.

1963. Elektronenoptische Untersuchungen über die Ascosporenbildung bei *Saccharomyces cerevisiae* unter cytologischem und cytogenetischem Aspekt. Arch. Mikrobiol. 46:308–320.

MICHELI, P. A.

1729. Nova plantarum genera juxta Tournefortii methodum disposita. Florence. 234 p.

MITTMANN, G.

1933. Kulturversuche mit Einsporstämmen und zytologische Untersuchungen in der Gattung *Ceratostomella*. Jahrb. wiss. Bot. 77:185–219.

MOORE, R. T.

1963. Fine structure of mycota. 1. Electron microscopy of the discomycete *Ascodesmis*. Nova Hedwigia 5:263–278.

1965. The ultrastructure of fungal cells, p. 95–118. *In* G. C. Ainsworth and A. S. Sussman [ed.] The fungi, Vol. I. Academic Press, New York.

MOTTA, J. J.

1969. Somatic nuclear division in *Armillaria mellea*. Mycologia 61:873–886.

OLIVE, L. S.

1953. The structure and behavior of fungus nuclei. Bot. Rev. 19:439–586.

1954. Taxonomic differentiation between *Ascobolus stercorarius* and *A. furfuraceus*. Mycologia 46:105–109.

OSO, B. A.

1969. Electron microscopy of ascus development in *Ascobolus*. Ann. Bot. 33:205–209.

PRUSSO, D. C., and K. WELLS

1967. *Sporobolomyces roseus*. I. Ultrastructure. Mycologia 59:337–348.

REEVES, F., JR.

1967. The fine structure of ascospore formation in *Pyronema domesticum*. Mycologia 59:1018–1033.

ROBERTSON, N. F.

1968. The growth process in fungi. Ann. Rev. Phytopath. 6:115–136.

ROBINOW, C. F., and C. E. CATEN

1969. Mitosis in *Aspergillus nidulans*. J. Cell Sci. 5:403–431.

ROBINOW, C. F., and J. MARAK

1966. A fiber apparatus in the nucleus of the yeast cell. J. Cell Biol. 29:129–151.

SCHRANTZ, J.-P.

1966. Contribution à l'étude de la formation de la paroi sporale chez *Pustularia cupularis* (L.) Fuck. C. r. Acad. Sci., Paris, Ser. D, 262:1212–1215.

1967. Présence d'un aster au cours des mitoses de l'asque et de la formation des ascospores chez l'Ascomycète *Pustularia cupularis* (L.) Fuck. C. r. Acad. Sci., Paris, Ser. D, 264:1274–1277.

1970*a*. Etude cytologique en microscopie optique et électronique, de quelques Ascomycètes. I. Le noyau. Rev. Cytol. Biol. veg. 33:1–67.

1970*b*. Etude cytologique en microscopie optique et électronique, de quelques Ascomycètes. II. La paroi. Rev. Cytol. Biol. veg. 33:111–168.

SCHWEIZER, G.

1932. Studien über die Kernverhältnisse im Archikarp von *Ascobolus furfuraceus* Pers. Ber. Deut. bot. Ges. 50A:14–23.

THIELKE, C.

1968. Membransysteme in meiotischen Basidien. Ber. Deut. bot. Ges. 81:183–186.

1969. Restitution der Kernmembran in postmeiotischen Basidien. Ber. Deut. bot. Ges. 81:315–316.

WELLS, K.

1964. The basidia of *Exidia nucleata*. II. Development. Amer. J. Bot. 51:360–370.

1965. Ultrastructural features of developing and mature basidia and basidiospores of *Schizophyllum commune*. Mycologia 57:236–261.

1970. Light and electron microscopic studies of *Ascobolus stercorarius*. I. Nuclear divisions in the ascus. Mycologia 62:761–790.

WELSFORD, E. J.
1907. Fertilization in *Ascobolus furfuraceus* Pers. New Phytol. 6:156–161.
WILSENACH, R., and M. KESSEL
1965. The role of lomasomes in wall formation in *Penicillium vermiculatum.* J. gen. Microbiol. 40:401–404.
WILSON, I. M.
1952. The ascogenous hyphae of *Pyronema confluens.* Ann. Bot. 16:321–339.
ZACHARIAH, K.
1970. Lattice bodies and filament bundles in apothecial cells of the fungus *Ascobolus.* Can. J. Bot. 48:1115–1118.
ZICKLER, D.
1970. Division spindle and centrosomal plaques during mitosis and meiosis in some Ascomycetes. Chromosoma 30:287–304.

PLATES

PLATE 1

Photomicrographs of the hymenium and asci.

A. Section of hymenium showing asci in longitudinal aspect. Fixed in 10 percent acrolein, embedded in glycol methacrylate, and stained with toluidine blue. × 350.

B. Longitudinal section of a diploid ascus prior to the formation of apical vacuoles. Fixed in 10 percent acrolein, embedded in glycol methacrylate, and stained with toluidine blue. × 2,100.

C. Longitudinal section of a diploid ascus with apical vacuoles. Fixed in 10 percent acrolein, embedded in glycol methacrylate, and stained with toluidine blue. × 2,100.

D. Distal portion of a diploid ascus with apical vacuoles. Photographed with Nomarski's interference-contrast optics. × 2,200.

E. Basal portion of an ascus containing a diploid nucleus. Squash mount fixed in Carnoy's solution and stained with acetocarmine. × 2,800.

AS, ascus; N, neucleus; NER, netted endoplasmic reticulum; nu, nucleolus; PA, paraphysis; PG, plug in pore of septum at base of ascus; va, apical vacuole; vb, basal vacuole.

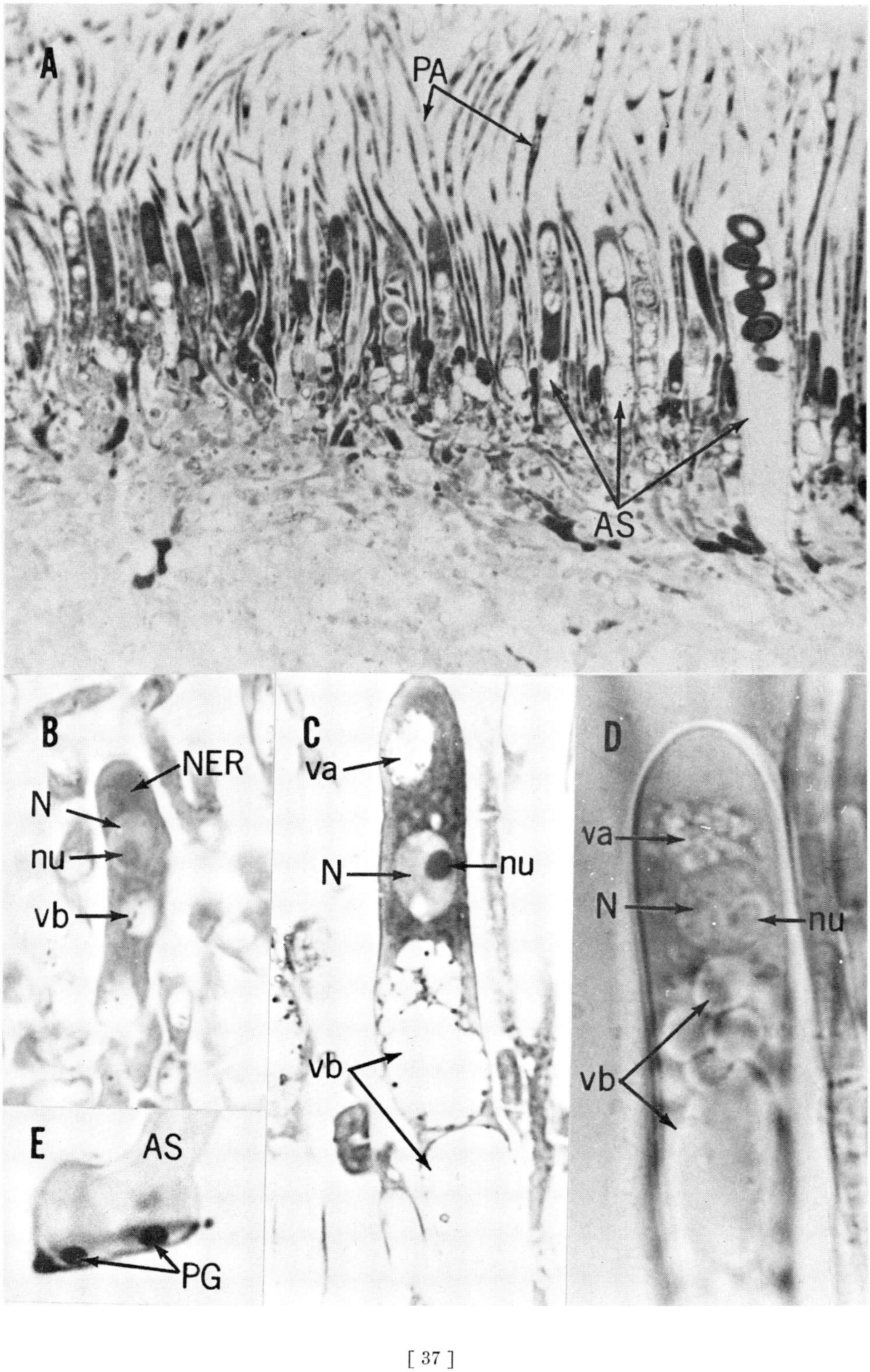
A
PA
AS
B
NER
N
nu
vb
C
va
N
nu
vb
D
va
N
nu
vb
E
AS
PG

PLATE 2

Photomicrographs of developing asci.

A. Longitudinal section of an ascus at the four-nucleate stage. Only three nuclei are visible. Fixed in 10 percent acrolein, embedded in glycol methacrylate, and stained with toluidine blue. × 2,000.

B. Apical portion of an ascus at the four-nucleate stage. Photographed with Nomarski's interference-contrast optics. × 2,100.

C. Longitudinal section of an ascus at the eight-nucleate stage. Only four nuclei are visible. Fixed in 10 percent acrolein, embedded in glycol methacrylate, and stained with toluidine blue. × 2,000.

D. Apical portion of an ascus with delimited, spherical ascospore initials. Photographed with Nomarski's interference-contrast optics. × 1,950.

E. Apical portion of an ascus photographed with Nomarski's interference-contrast optics. The endospore layer of the ascospore wall is well developed. × 1,900.

F. Apical portion of an ascus photographed with Nomarski's interference-contrast optics. The ascospores have a distinct perisporal sac. × 1,800.

ASP, ascospore; EP, epiplasm; L, lipid globule; N, nucleus; pl, perisporal sac; V, vacuole; va, apical vacuole; vb, basal vacuole; WS, wall of ascospore.

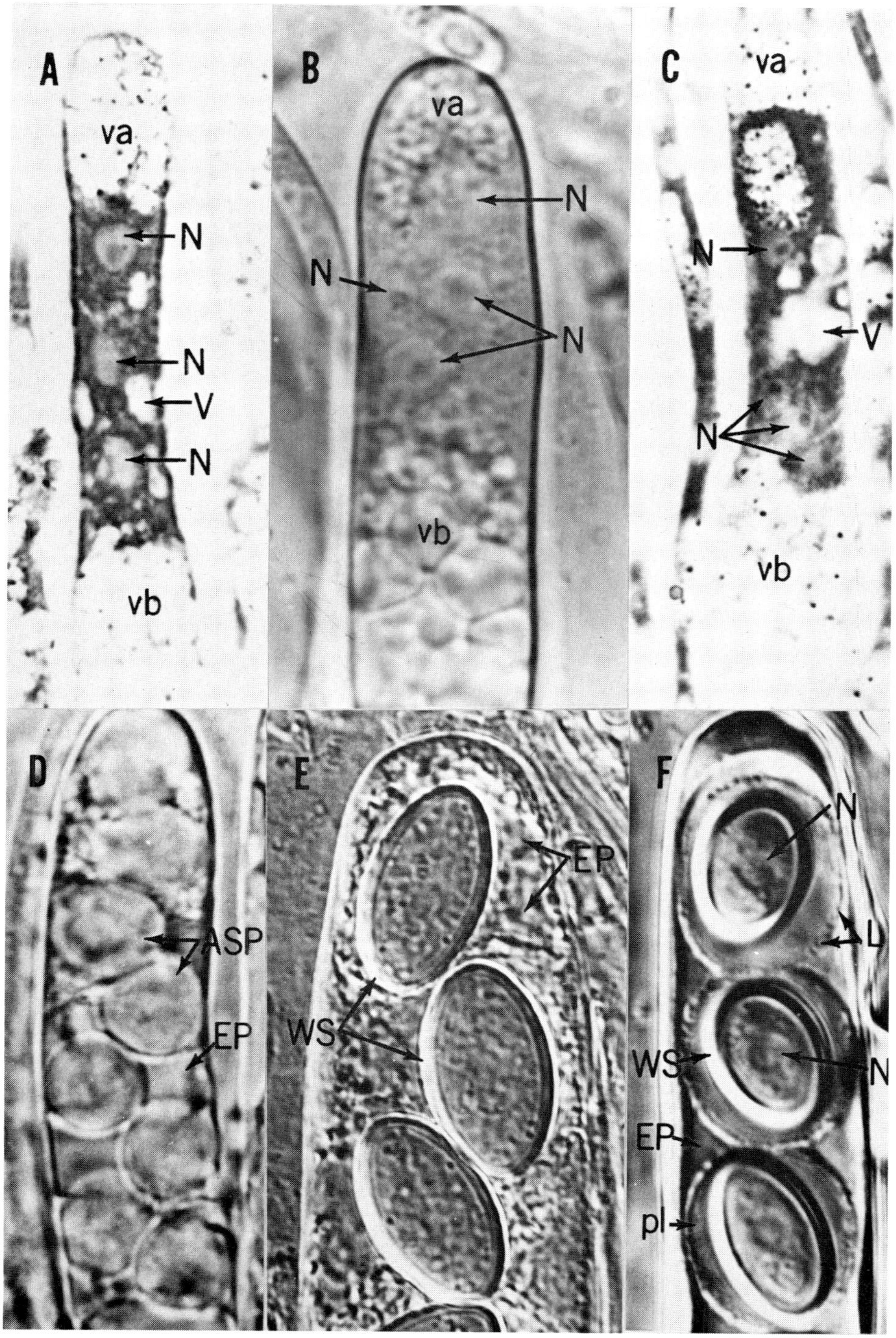
A
va
N
N
V
N
vb
B
va
N
N
N
vb
C
va
N
V
N
vb
D
ASP
EP
E
EP
WS
F
N
L
WS
N
EP
pl

PLATE 3

Photomicrographs of asci and ascospores.

A. Apical portion of an ascus photographed with Nomarski's interference-contrast optics. The ascospores have a distinct, unilateral perisporal sac. × 1,600.

B. Apical portion of an ascus containing nearly mature spores. Photographed with Nomarski's interference-contrast optics. × 1,900.

C. Portion of an ascus containing mature ascospores. Photographed with Nomarski's interference-contrast optics. × 1,200.

D. Apical portion of an ascus. The ascospores have a distinct perisporal sac. Fixed in 50 percent ethanol and stained with Sudan black B. × 1,800.

E. Tip of ascus following ascospore discharge. Photographed with Nomarski's interference-contrast optics. × 1,900.

F. Two ascospores. The perisporal sac is intact around the lower spore, but it has been torn from the upper spore. × 1,900.

EP, epiplasm; L, lipid globule; OP, operculum; PGR, pigment granules in perisporal sac; pl, perisporal sac; WS, wall of ascospore.

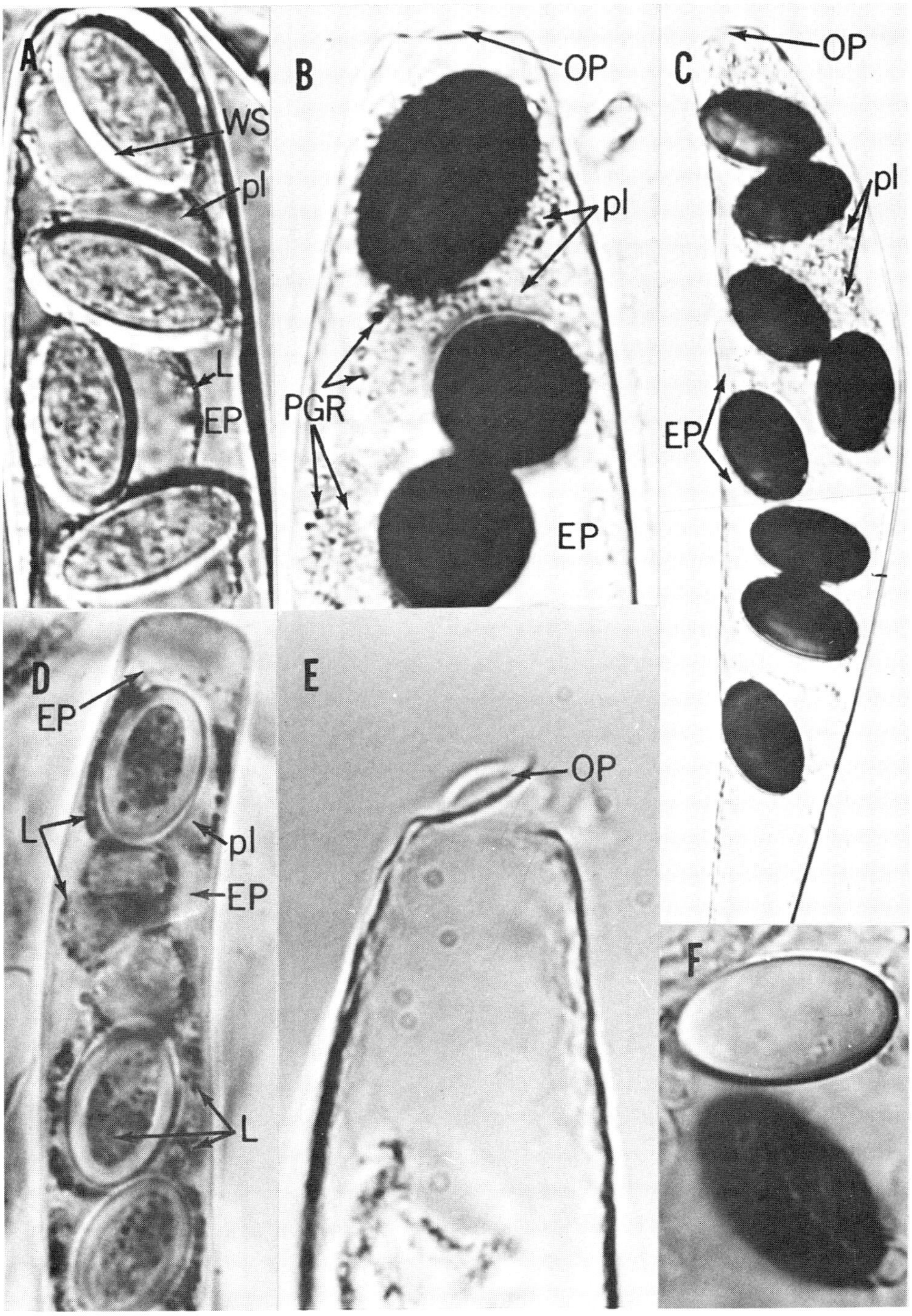
A
WS
pl
L
EP
B
OP
pl
PGR
EP
C
OP
pl
EP
D
EP
L
pl
EP
L
E
OP
F

PLATE 4

Electron micrographs of a crozier and an ascus prior to karyogamy. Fixed in 1.5 percent $KMnO_4$ and stained with lead citrate.

A. Longitudinal section of a crozier prior to conjugate nuclear division and segmentation. × 8,800.

B. Longitudinal section of an ascus containing a dikaryon. × 12,350.

AW, ascus wall; CB, location of electron-dense substance adjacent to prefusion and fusion nuclei fixed in glutaraldehyde-OsO_4 (see Wells, 1970); CW, wall of crozier; dm, electron-dense portion of gelatinous matrix surrounding asci and paraphyses; ER, endoplasmic reticulum; m, mitochondrion; N, nucleus; PM, plasma membrane; V, vacuole.

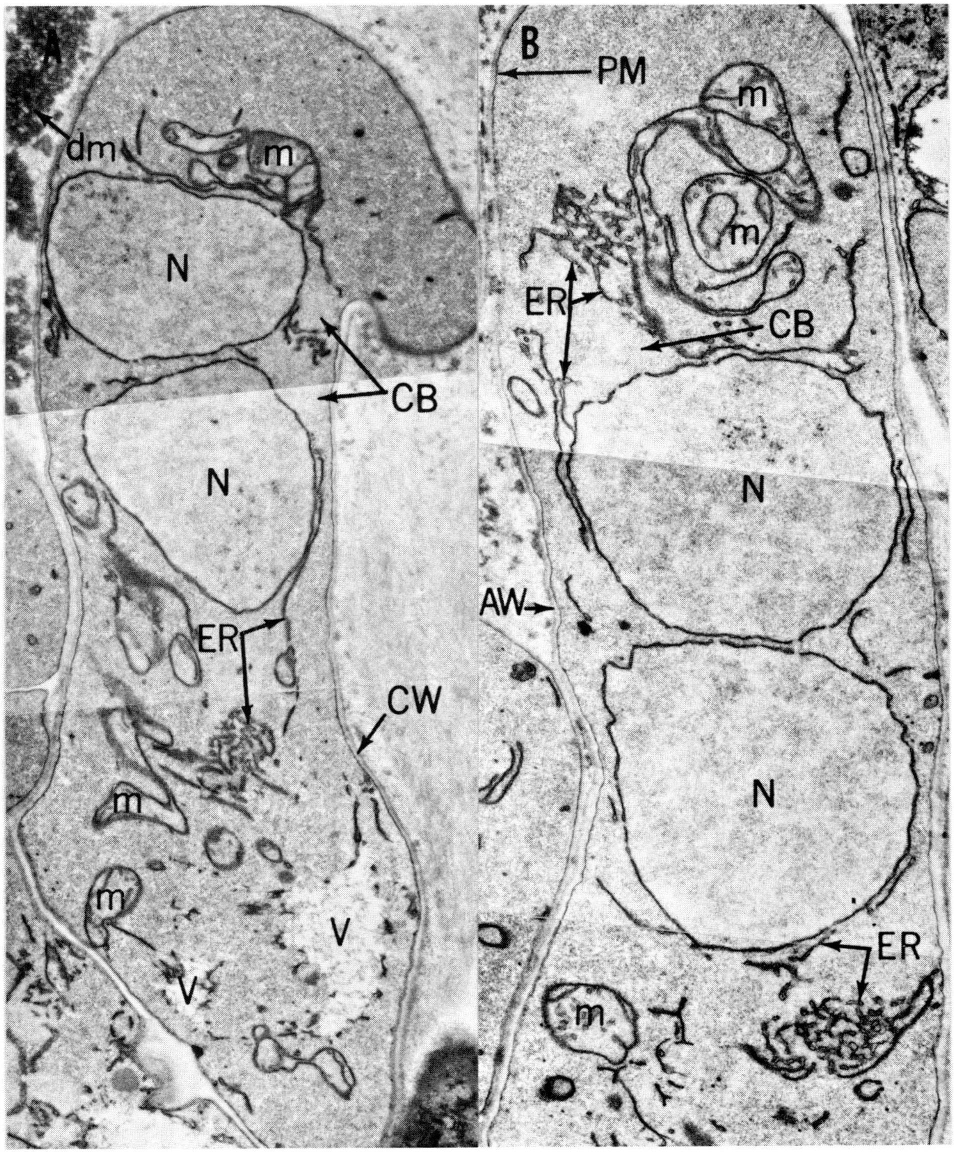
A
dm
m
N
CB
N
ER
CW
m
m
V
V
B
PM
m
m
ER
CB
N
AW
N
ER
m

PLATE 5

Electron micrographs of asci containing diploid nuclei. Fixed in 1.5 percent $KMnO_4$ and stained with lead citrate.

A. Apical portion of an ascus with diploid nucleus prior to appearance of the netted endoplasmic reticulum. A portion of the diploid nucleus appears separate in this electron micrograph; however, adjacent thin sections show that it is only a projection. × 12,000.

B. Apical portion of an ascus at the diploid stage containing netted endoplasmic reticulum and a few apical vacuoles. × 5,500.

AW, ascus wall; CB, location of electron-dense substance adjacent to prefusion and fusion nuclei fixed in glutaraldehyde-OsO_4 (see Wells, 1970); ER, endoplasmic reticulum; m, mitochondrion; N, nucleus; NER, netted endoplasmic reticulum; PA, paraphysis; PM, plasma membrane; va, apical vacuole; vb, basal vacuole.

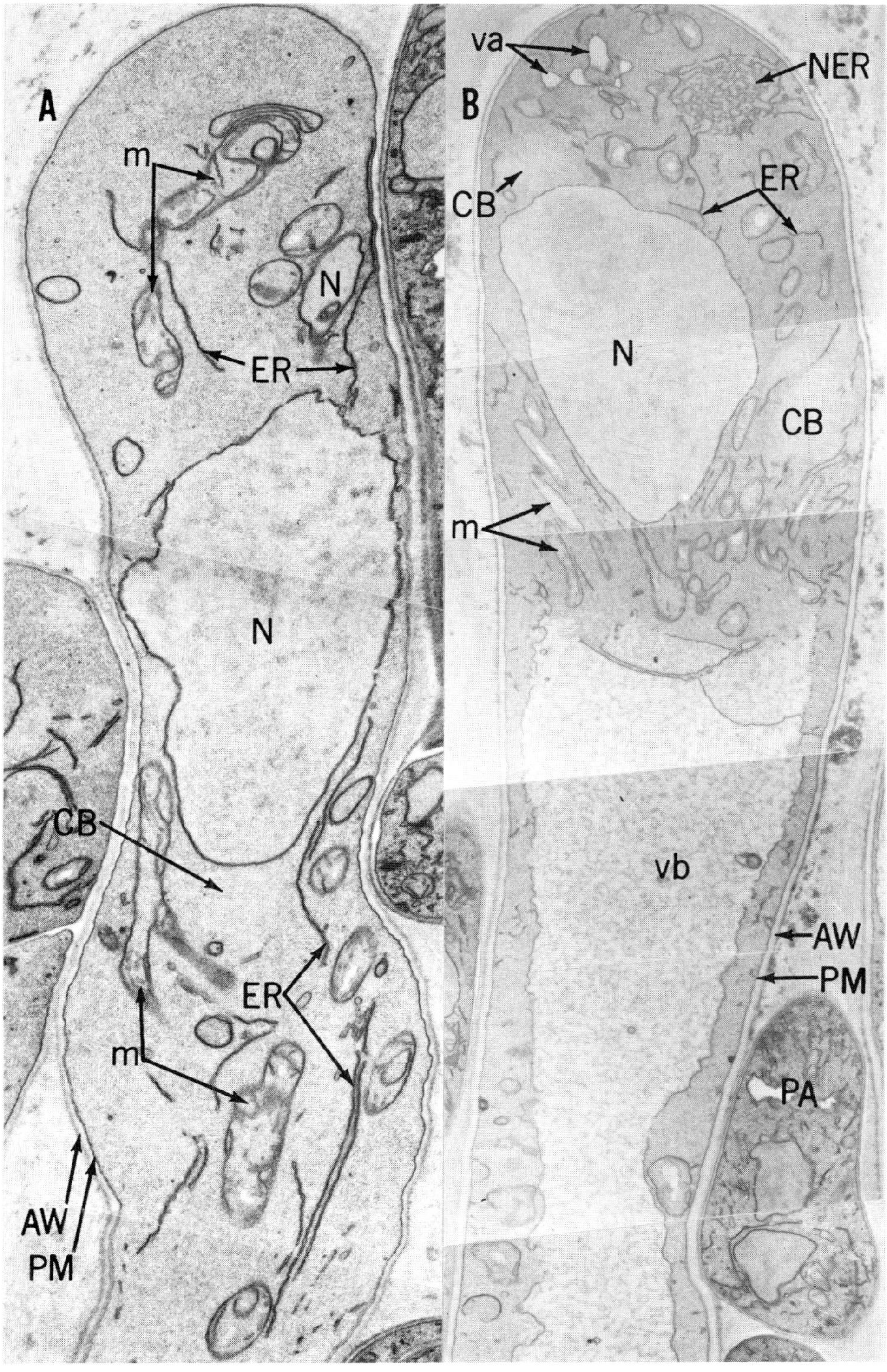
A
m
N
ER
N
CB
ER
m
AW
PM
B
va
NER
CB
ER
N
CB
m
vb
AW
PM
PA

PLATE 6

Electron micrographs illustrating the netted endoplasmic reticulum above the diploid nucleus.

A. Section of the apical portion of an ascus. Fixed in 6.5 percent glutaraldehyde and 1 percent OsO_4 and stained with lead citrate. × 15,200.

B. Portion of cytoplasm in the apical region of an ascus prior to the formation of the apical vacuoles. Fixed in 1.5 percent $KMnO_4$ and stained with lead citrate. × 26,000.

CB, location of electron-dense substance adjacent to prefusion and fusion nuclei fixed in glutaraldehyde-OsO_4 (see Wells, 1970) ; ER, endoplasmic reticulum; m, mitochondrion; N, nucleus; NER, netted endoplasmic reticulum.

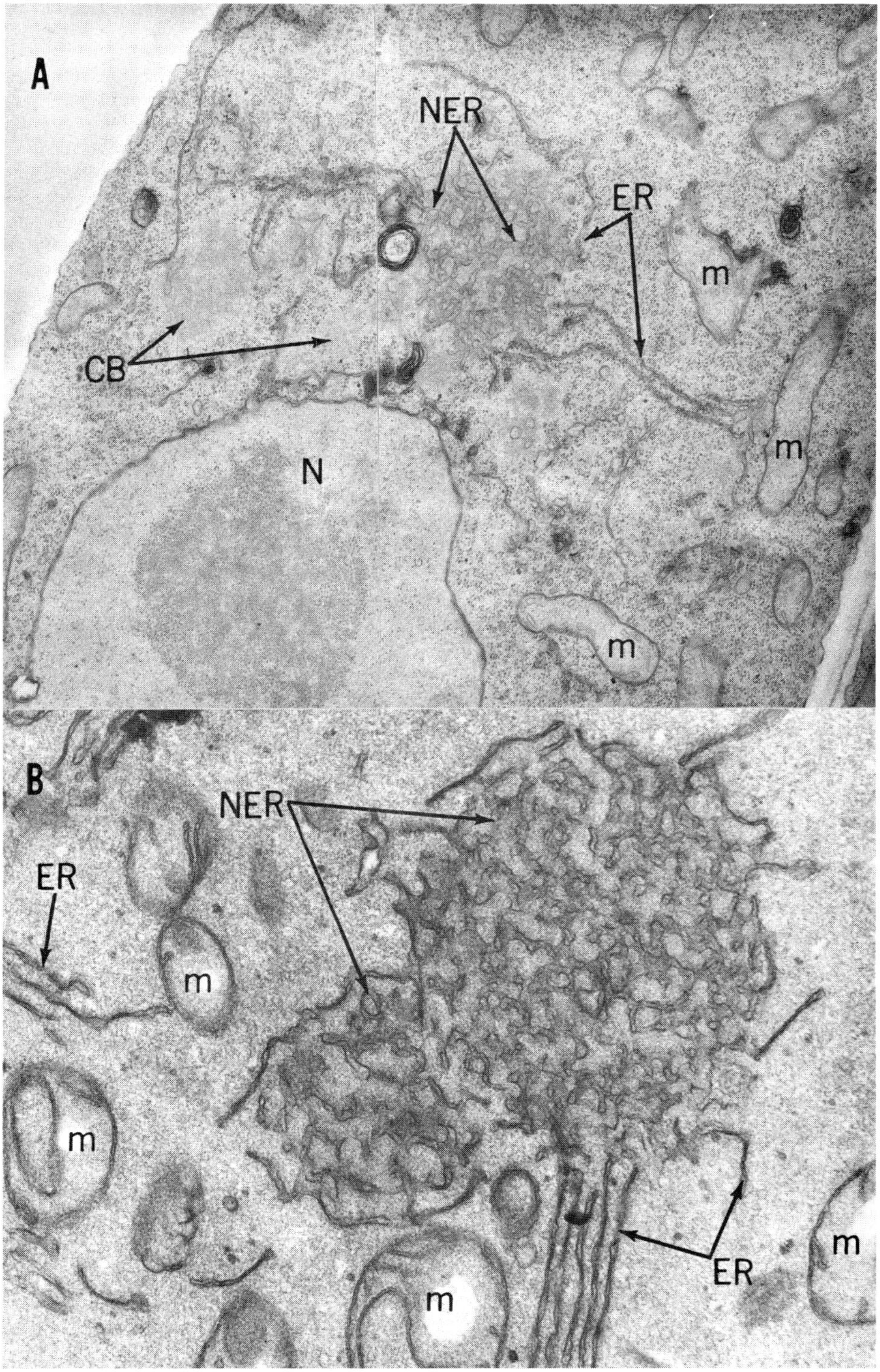
A
NER
ER
m
CB
m
N
m
B
NER
ER
m
m
m
ER
m

PLATE 7

Electron micrographs of portions of asci with diploid nuclei. Fixed in 1.5 percent $KMnO_4$ and stained with lead citrate.

A. Longitudinal section of distal portion of ascus. × 4,200.

B. Longitudinal section of distal portion of ascus. × 5,500.

C. Section of basal portion of ascus and a portion of attached ascogenous hypha. × 5,800.

a, apical vesicle; AS, ascus; AW, ascus wall; CB, location of electron-dense substance adjacent to prefusion and fusion nuclei fixed in glutaraldehyde-OsO_4 (see Wells, 1970); dm, electron-dense portion of matrix surrounding asci and paraphyses; ER, endoplasmic reticulum; m, mitochondrion; N, nucleus; PA, paraphysis; PG, plug in pore of septum at base of ascus; se, septum; ST, stalk cell; T, tonoplast; TC, terminal cell; va, apical vacuole; vb, basal vacuole.

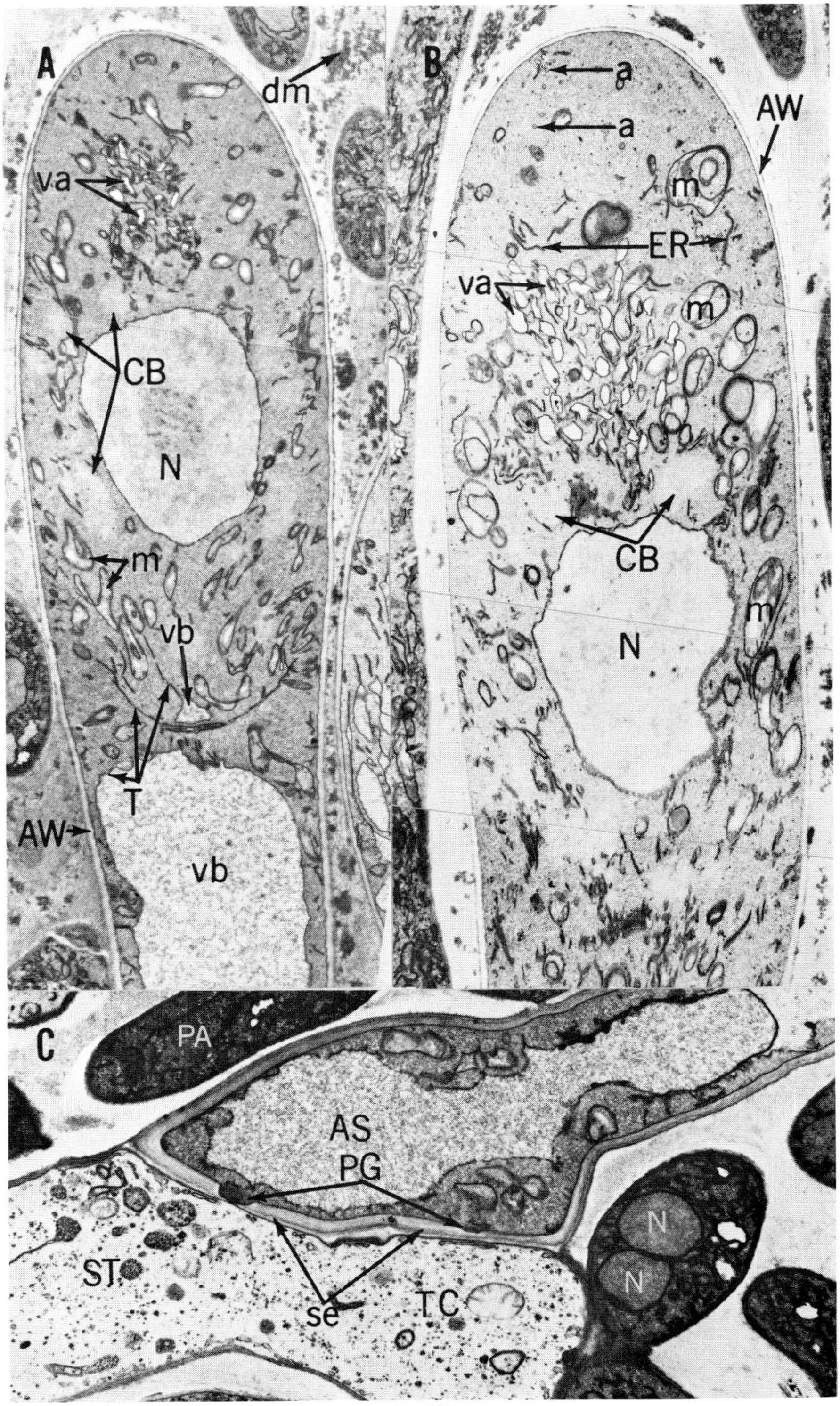
A
dm
va
CB
N
m
vb
T
AW
vb
B
a
a
AW
m
ER
va
m
CB
N
m
C
PA
AS
PG
ST
se
TC
N
N

PLATE 8

Electron micrographs of the basal septa of asci containing diploid nuclei.

A. Cross section of septum between an ascus and a stalk cell. Fixed in 1.5 percent $KMnO_4$ and stained with lead citrate. × 22,900.

B. Cross section of portions of the septa between the ascus and the stalk cell and between the terminal cell and the ascus. Fixed in 1 percent glutaraldehyde and 2 percent OsO_4 and stained with barium permanganate and lead citrate. × 26,900.

AS, ascus; AW, ascus wall; CM, concentric membrane complex; ER, endoplasmic reticulum; g, glycogen; lo, lomasome; PG, plug in pore of septum at base of ascus; se, septum; ST, stalk cell; TC, terminal cell.

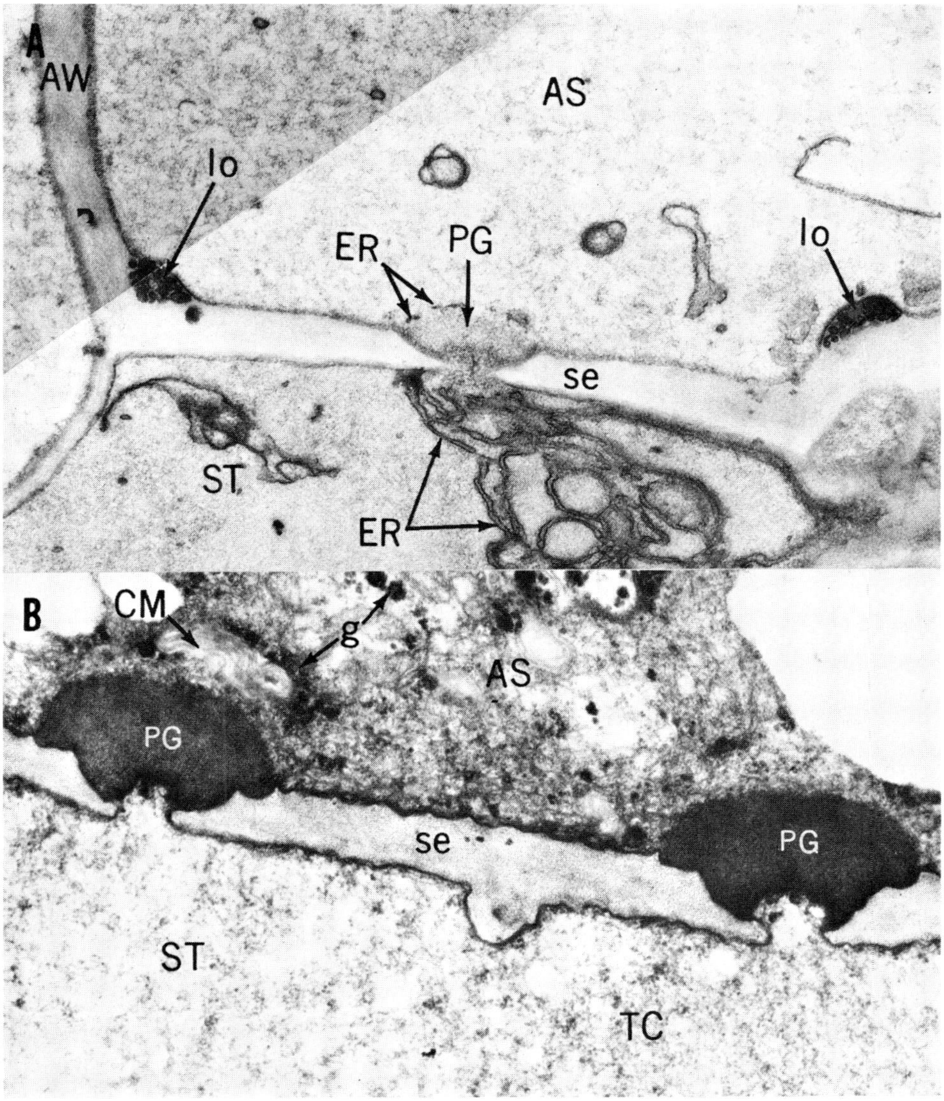
A
AW
AS
lo
ER
PG
lo
se
ST
ER
B
CM
g
AS
PG
se
PG
ST
TC

PLATE 9

Electron micrograph of a longitudinal section of the apical portion of an ascus at the four-nucleate stage prior to the initiation of the double ascospore-delimiting membranes. Fixed in 1.5 percent $KMnO_4$ and stained with lead citrate. $\times$ 8,000.

a, apical vesicle; AW, ascus wall; dm, electron-dense portion of gelatinous matrix surrounding asci and paraphyses; ER, endoplasmic reticulum; g, glycogen; m, mitochondrion; N, nucleus; PA, paraphysis; PM, plasma membrane; va, apical vacuole.

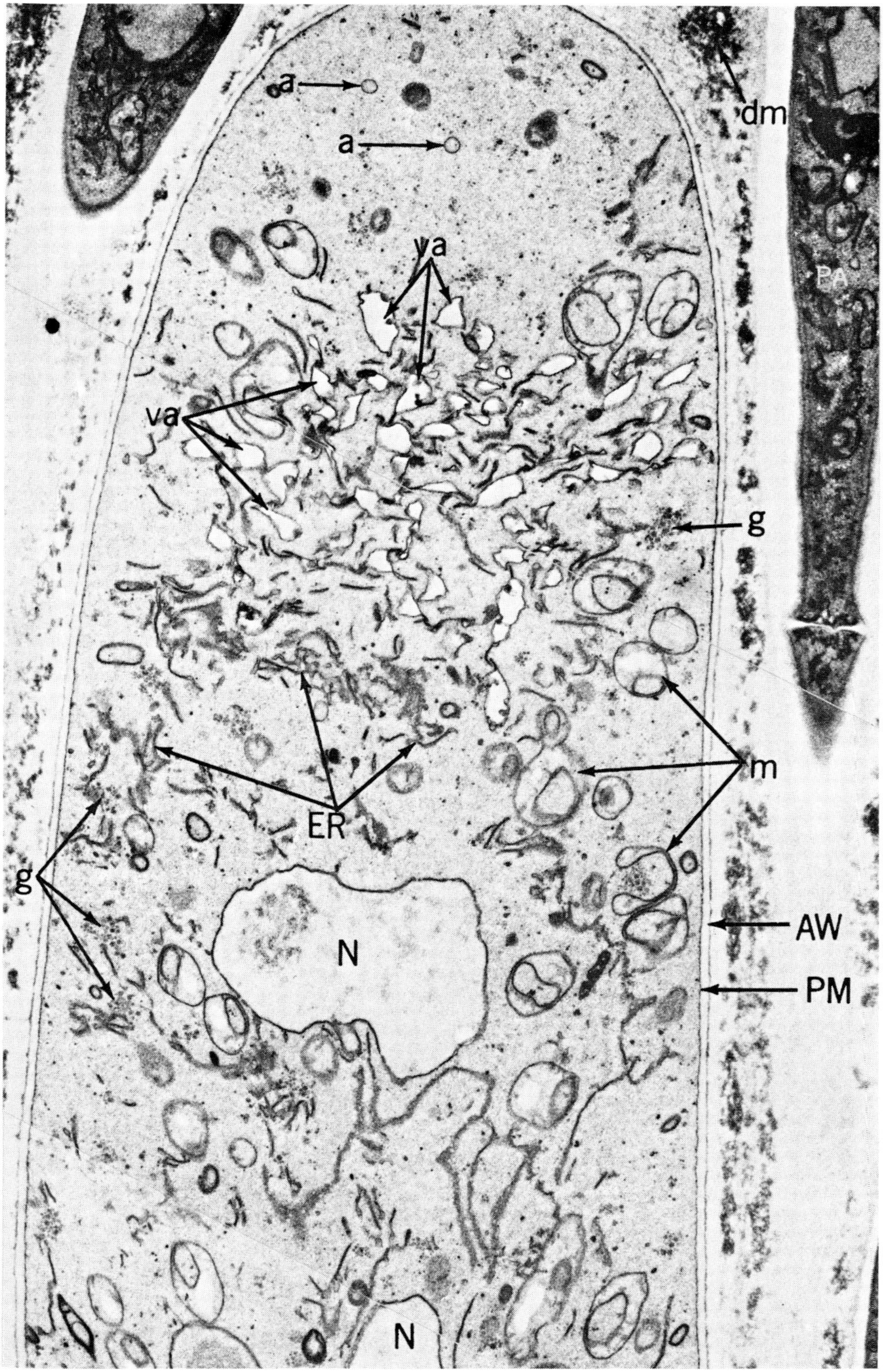
a
a
dm
va
va
PA
g
ER
m
g
AW
PM
N
N

PLATE 10

Electron micrographs of longitudinal sections of the distal portion of asci in the four-nucleate stage. Only three nuclei are visible in the electron micrographs of each ascus. Fixed in 1.5 percent $KMnO_4$ and stained with lead citrate.

A. Section illustrating ascus at four-nucleate stage prior to the appearance of the double ascospore delimiting membranes. × 5,000.

B. Section illustrating ascus at four-nucleate stage after the initiation of the double ascospore-delimiting membranes. × 4,100.

AD, double ascospore-delimiting membrane; AW, ascus wall; ER, endoplasmic reticulum; g, glycogen; m, mitochondrion; N, nucleus; nb, nuclear bleb; va, apical vacuole; vb, basal vacuole.

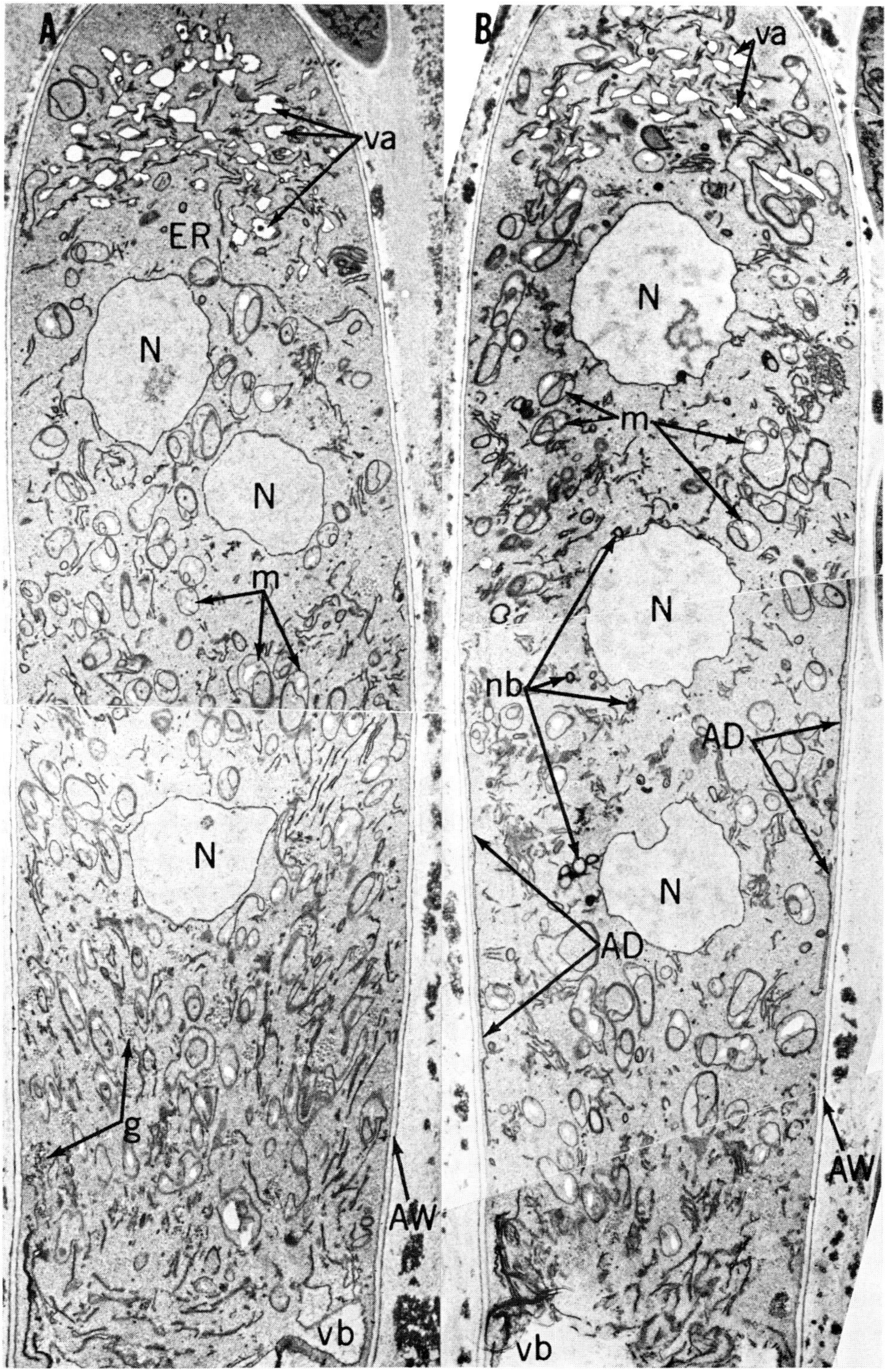
A
va
ER
N
N
m
N
g
AW
vb
B
va
N
m
N
nb
AD
N
AD
AW
vb

PLATE 11

Electron micrographs illustrating nuclear blebbing.

A. A portion of a haploid nucleus at the four-nucleate stage showing a nuclear bleb and, presumably, an earlier stage in the formation of a nuclear bleb. Fixed in 1.5 percent $KMnO_4$ and stained with lead citrate. × 53,650.

B. A portion of a haploid nucleus at the four-nucleate stage illustrating, presumably, an intermediate stage in nuclear bleb formation. Fixed in a mixture of 1.5 percent glutaraldehyde and 1.5 percent $KMnO_4$ and postfixed in 1.5 percent $KMnO_4$. Sections stained with lead citrate. × 42,800.

ER, endoplasmic reticulum; m, mitochondrion; N, nucleus; nb, nuclear bleb; NE, nuclear envelope.

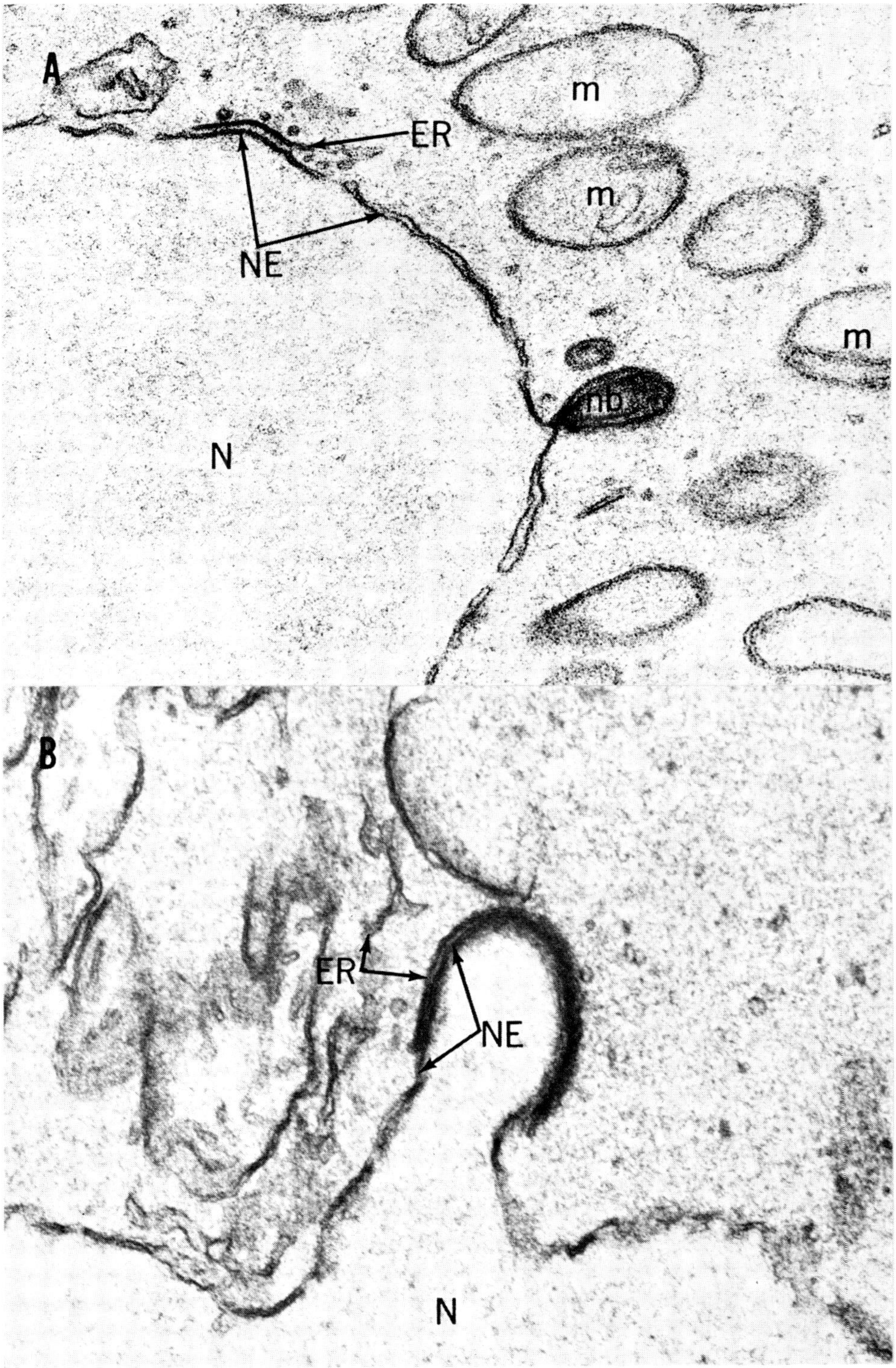
A
m
ER
m
NE
m
nb
N
B
ER
NE
N

PLATE 12

Electron micrographs illustrating the double ascospore-delimiting membranes. Fixed in 1.5 percent $KMnO_4$ and stained with lead citrate.

A. Peripheral portion of a four-nucleate ascus showing a small segment of a double ascospore-delimiting membrane. × 16,800.

B. A different portion of the ascus shown in A. × 14,100.

C. A portion of an ascus at the four-nucleate stage. × 19,400.

AD, double ascospore-delimiting membranes; AW, ascus wall; ER, endoplasmic reticulum; g, glycogen; m, mitochondrion; N, nucleus; nb, nuclear bleb; PM, plasma membrane.

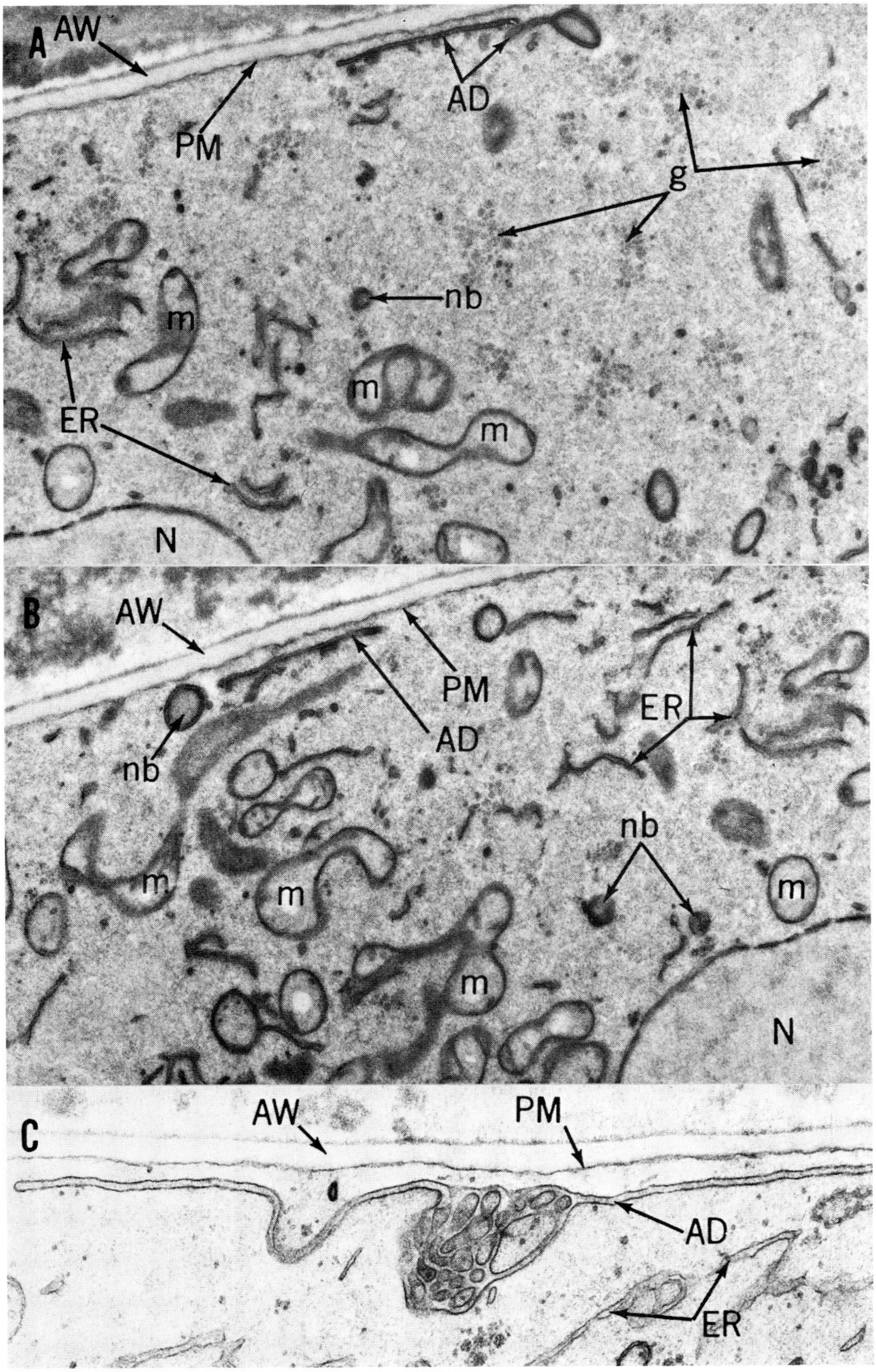
A
AW
PM
AD
g
nb
m
ER
N
B
AW
PM
AD
ER
nb
m
N
C
AW
PM
AD
ER

PLATE 13

Electron micrograph of a portion of an ascus at the eight-nucleate stage with inwardly developing double ascospore-delimiting membranes. Fixed in 1 percent glutaraldehyde and 2 percent OsO_4 and stained with barium permanganate and lead citrate. × 10,000.

AD, double ascospore-delimiting membranes; AW, ascus wall; CP, centriolar plaque; g, glycogen; L, lipid globule; N, nucleus; PA, paraphysis; PM, plasma membrane; V, vacuole; WB, Woronin body.

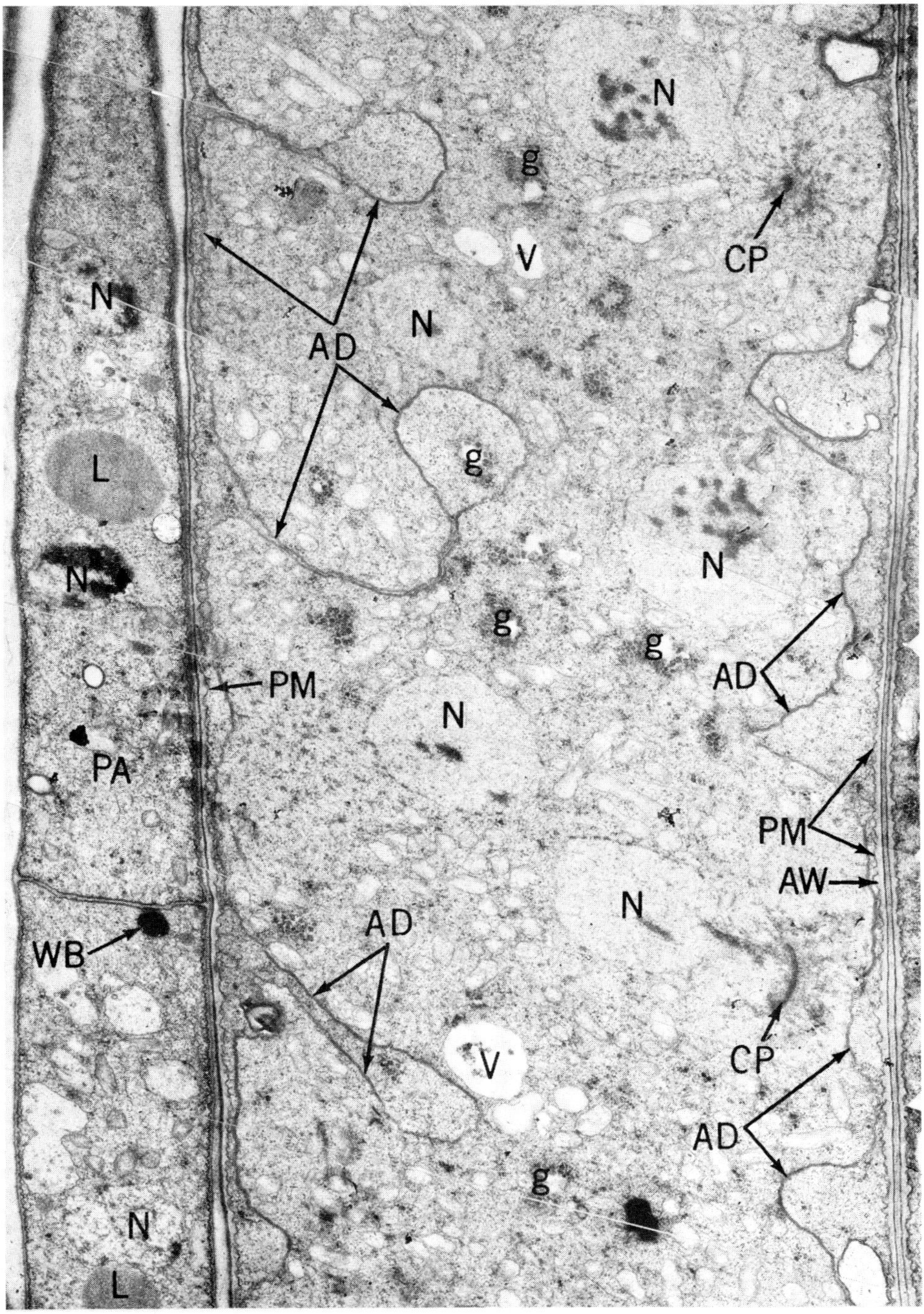
N
g
CP
V
N
AD
N
L
g
N
N
g
g
AD
PM
N
PA
PM
AW
N
WB
AD
CP
V
AD
g
N
L

PLATE 14

Electron micrograph of distal portion of an ascus with partially delimited ascospores. In several regions the double ascospore-delimiting membranes have not completely differentiated the sporoplasm and epiplasm. Fixed in 1.5 percent $KMnO_4$ and stained with lead citrate. × 8,000.

AD, double ascospore-delimiting membranes; AW, ascus wall; EP, epiplasm; g, glycogen; m, mitochondrion; N, nucleus; nb, nuclear bleb; PM, plasma membrane; SP, sporoplasm; va, apical vacuole.

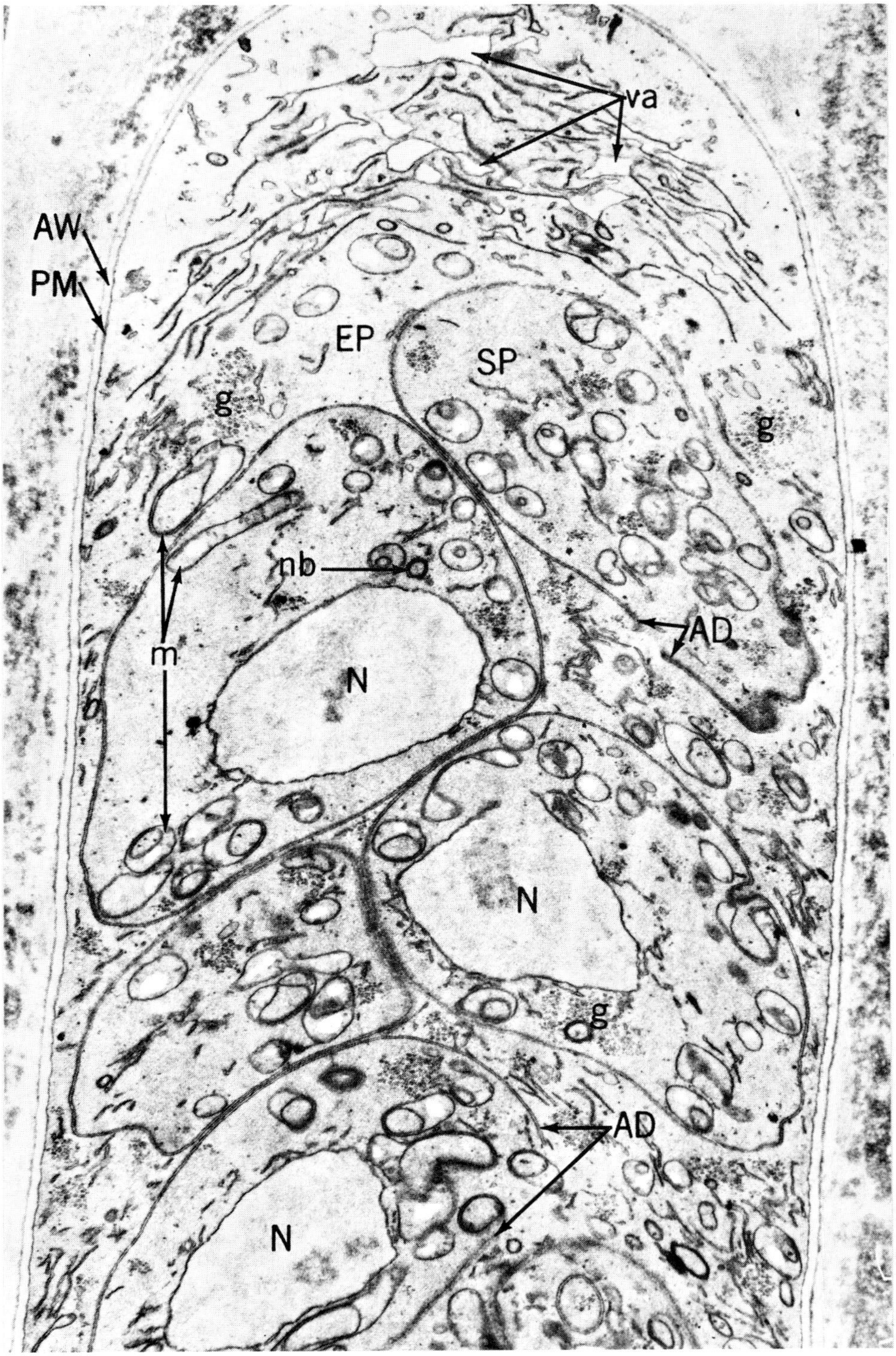
va
AW
PM
EP
SP
g
g
nb
m
AD
N
N
g
AD
N

PLATE 15

Electron micrograph of the sporogenous region of an ascus in the terminal stages of ascospore delimitation. Fixed in 1 percent glutaraldehyde and 2 percent OsO_4 and stained with uranyl acetate and lead citrate. × 9,500.

AD, double ascospore-delimiting membranes; AW, ascus wall; CM, concentric membrane complex; CP, centriolar plaque; EP, epiplasm; g, glycogen; N, nucleus; SP, sporoplasm; V, vacuole.

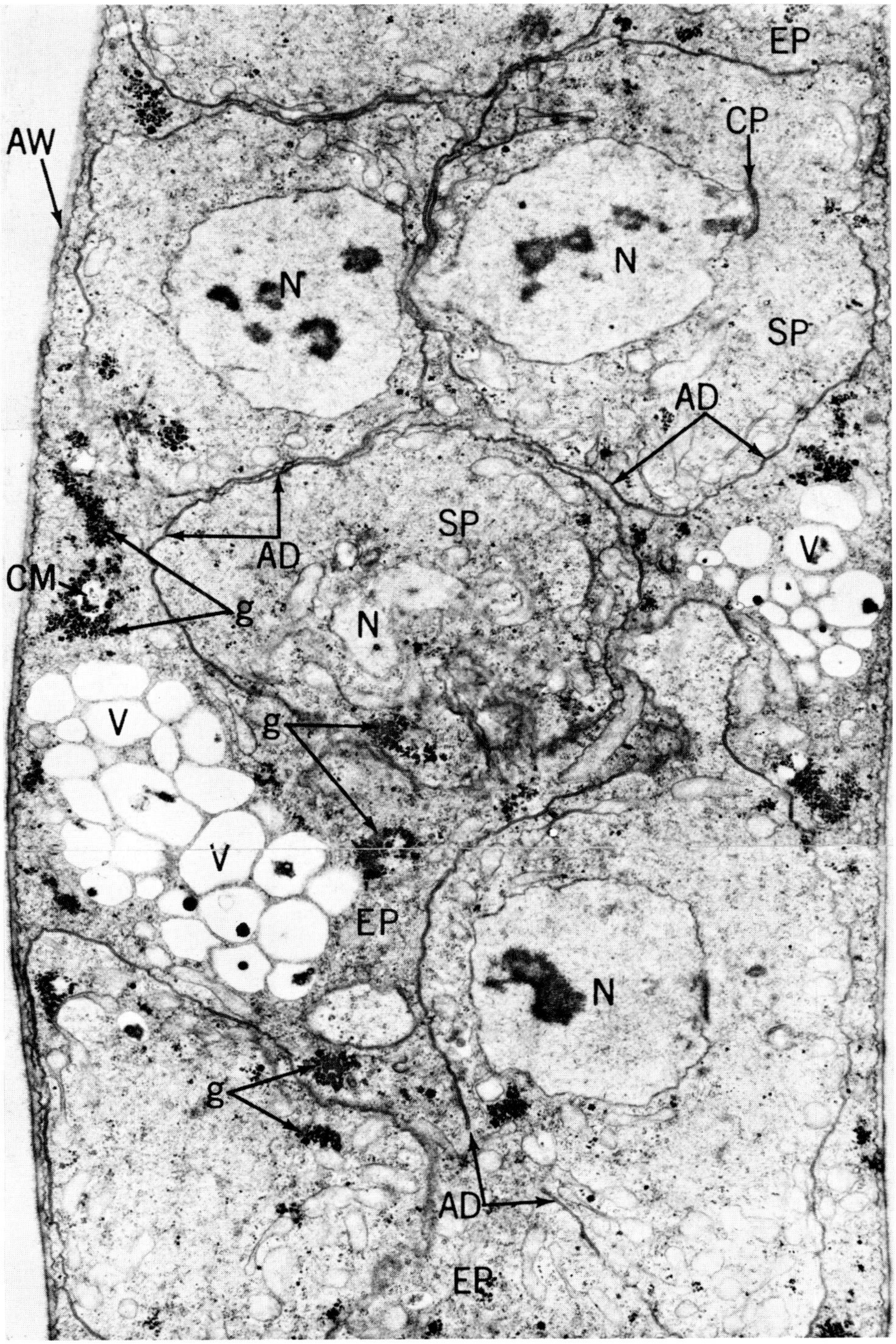
EP
CP
AW
N
N
SP
AD
SP
AD
V
CM
g
N
g
V
V
EP
N
g
AD
EP

PLATE 16

Electron micrograph of a portion of the sporogenous region of an ascus with delimited ascospores. Fixed in 1.5 percent $KMnO_4$ and stained with lead citrate. × 9,700.

AD, double ascospore-delimiting membranes; AW, ascus wall; EP, epiplasm; ER, endoplasmic reticulum; g, glycogen; m, mitochondrion; N, nucleus; nb, nuclear bleb; PM, plasma membrane; SP, sporoplasm; va, apical vacuole.

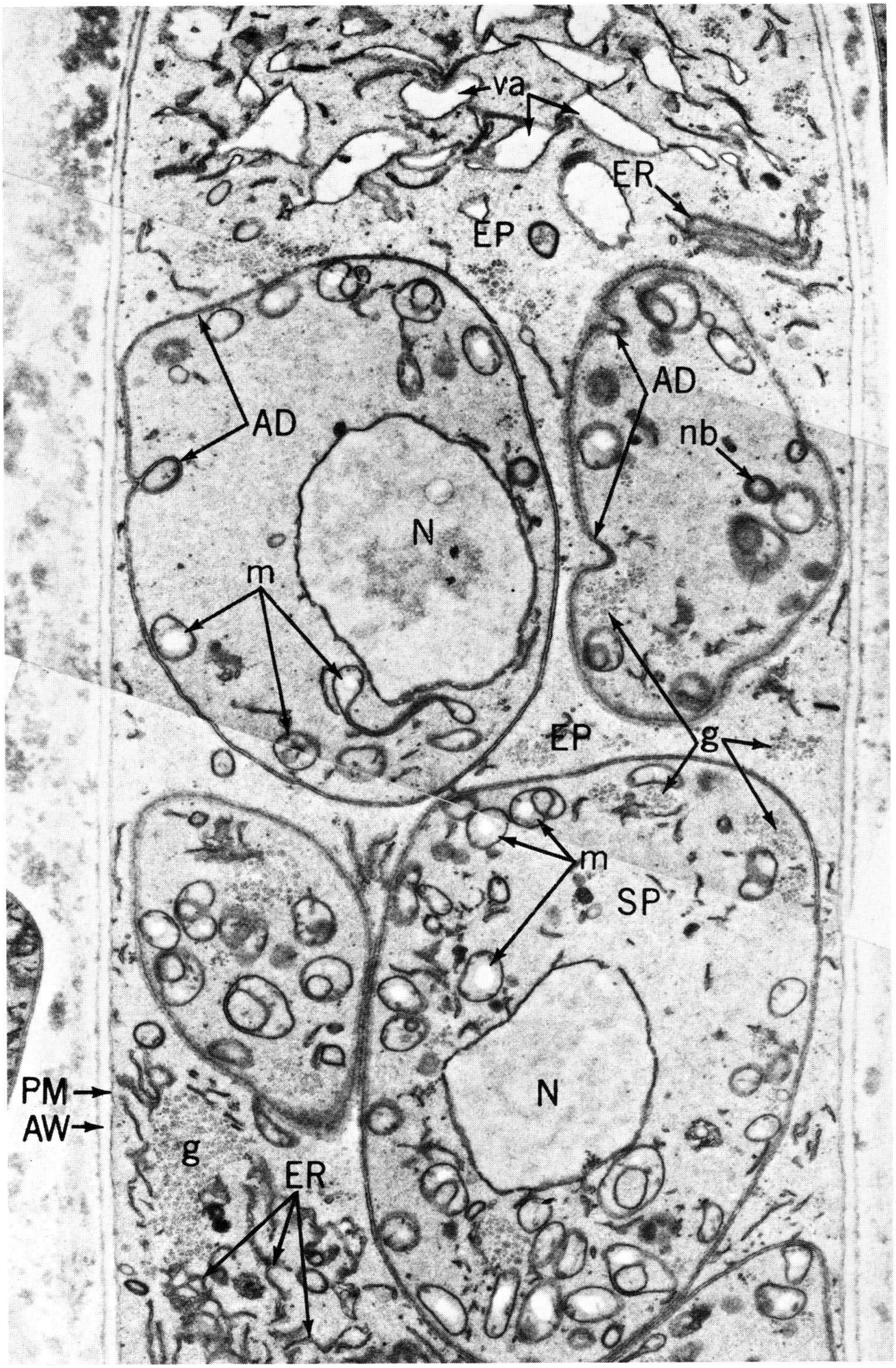
va
ER
EP
AD
AD
nb
N
m
EP
g
m
SP
N
PM
AW
g
ER

PLATE 17

Electron micrograph of a portion of a longitudinal section of an ascus showing a completely delimited ascospore. Fixed in 1 percent glutaraldehyde and 2 percent OsO_4 and stained with barium permanganate and lead citrate. $\times$ 13,700.

AD, double ascospore-delimiting membranes; AW, ascus wall; CM, concentric membrane complex; EP, epiplasm; g, glycogen; m, mitochondrion; N, nucleus; PM, plasma membrane; SP, sporoplasm.

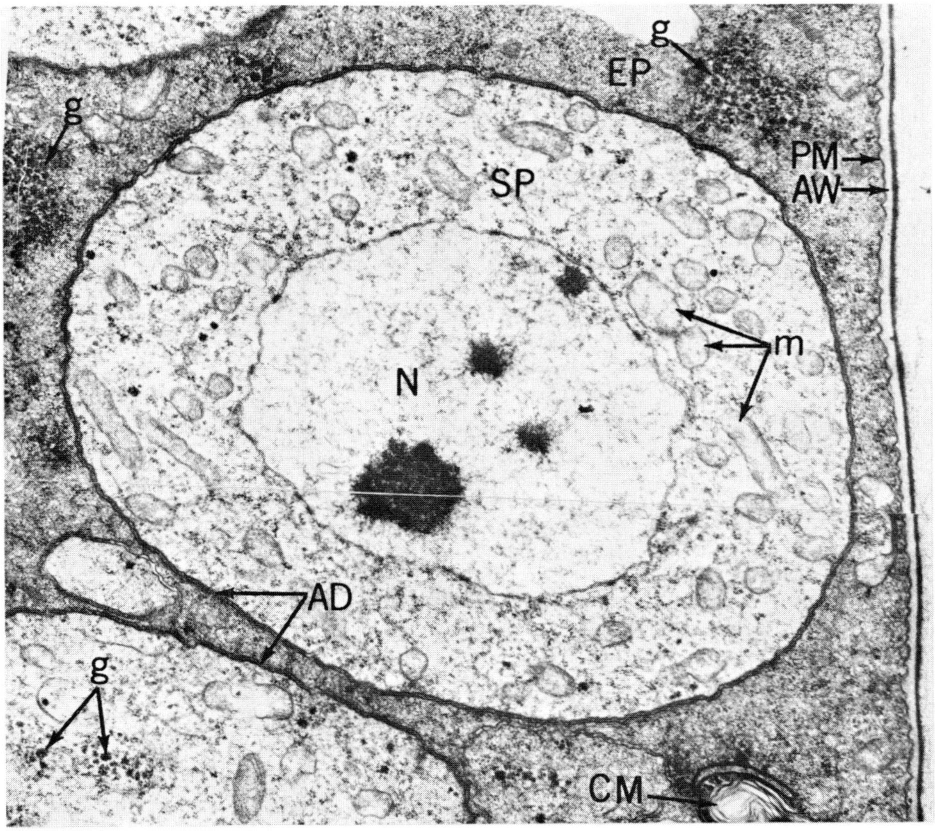
g
EP
g
PM
AW
SP
m
N
AD
g
CM

PLATE 18

Electron micrograph of a longitudinal section of the distal portion of an ascus following delimitation and realignment of the ascospores. Fixed in 1.5 percent $KMnO_4$ and stained with lead citrate. × 8,800.

a, apical vesicle; AW, ascus wall; dm, electron-dense portion of gelatinous matrix surrounding asci and paraphyses; EP, epiplasm; ER, endoplasmic reticulum; g, glycogen; iad, inner ascospore-delimiting membrane; m, mitochondrion; N, nucleus; oad, outer ascospore-delimiting membrane; PM, plasma membrane; SP, sporoplasm; ve, vacuole in epiplasm.

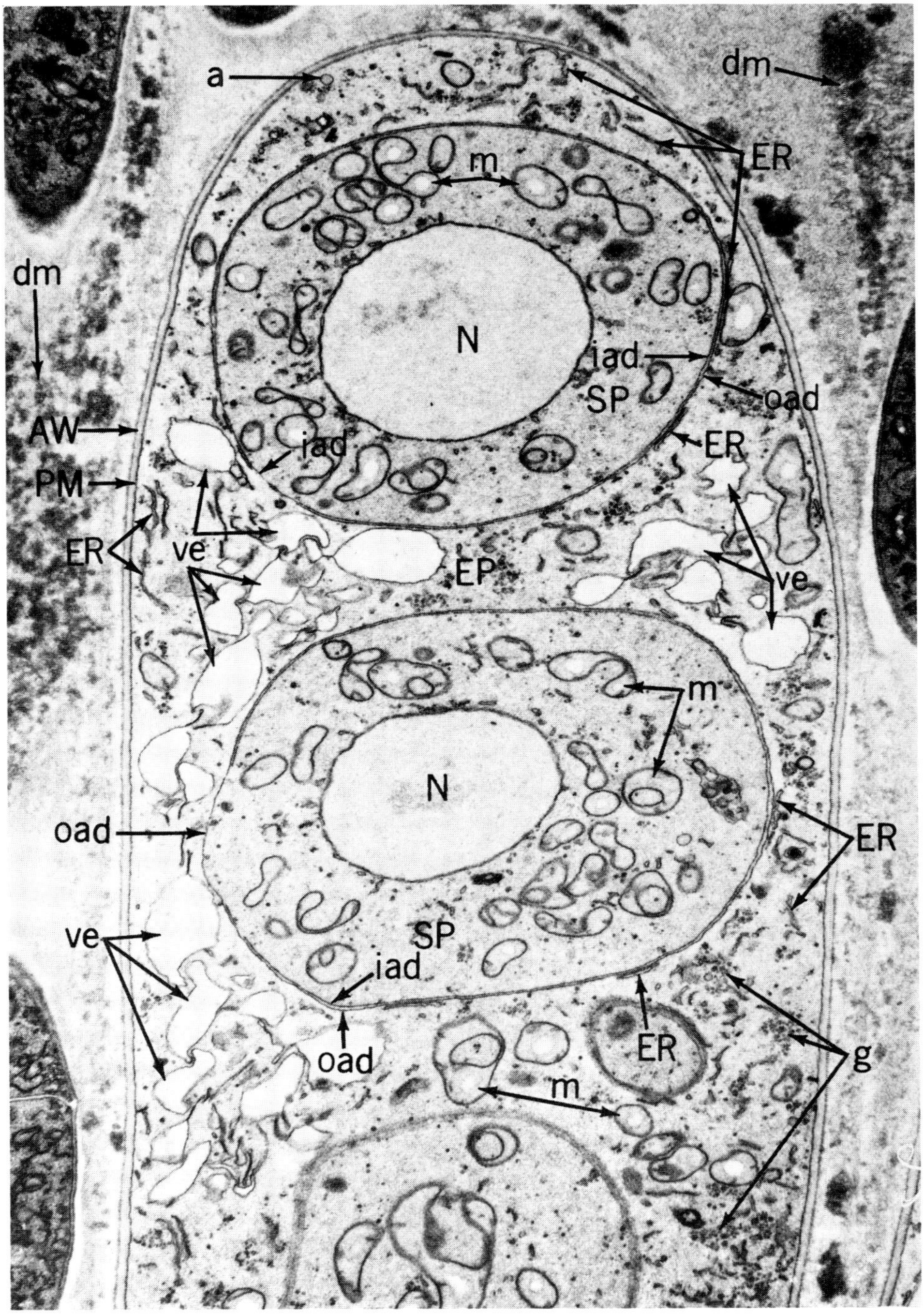
a
dm
ER
m
dm
N
iad
SP
oad
AW
iad
ER
PM
ER
ve
EP
ve
m
N
oad
ER
ve
SP
iad
oad
ER
g
m

PLATE 19

Electron micrograph of a longitudinal section of apical portion of an ascus with immature spores. The apical ascospore was sectioned longitudinally, whereas the median spore was sectioned obliquely. Fixed in 1.5 percent $KMnO_4$ and stained with lead citrate. × 8,500.

a, apical vesicle; AW, ascus wall; en, endospore; EP, epiplasm; ER, endoplasmic reticulum; im, investing membrane; m, mitochondrion; N, nucleus; PM, plasma membrane; SP, sporoplasm; ve, vacuole in epiplasm. Unlabelled arrow indicates region that probably contained fibrous or fine-grained structures, both of which are preserved in glutaraldehyde-OsO_4 (see pl. 27, A, B).

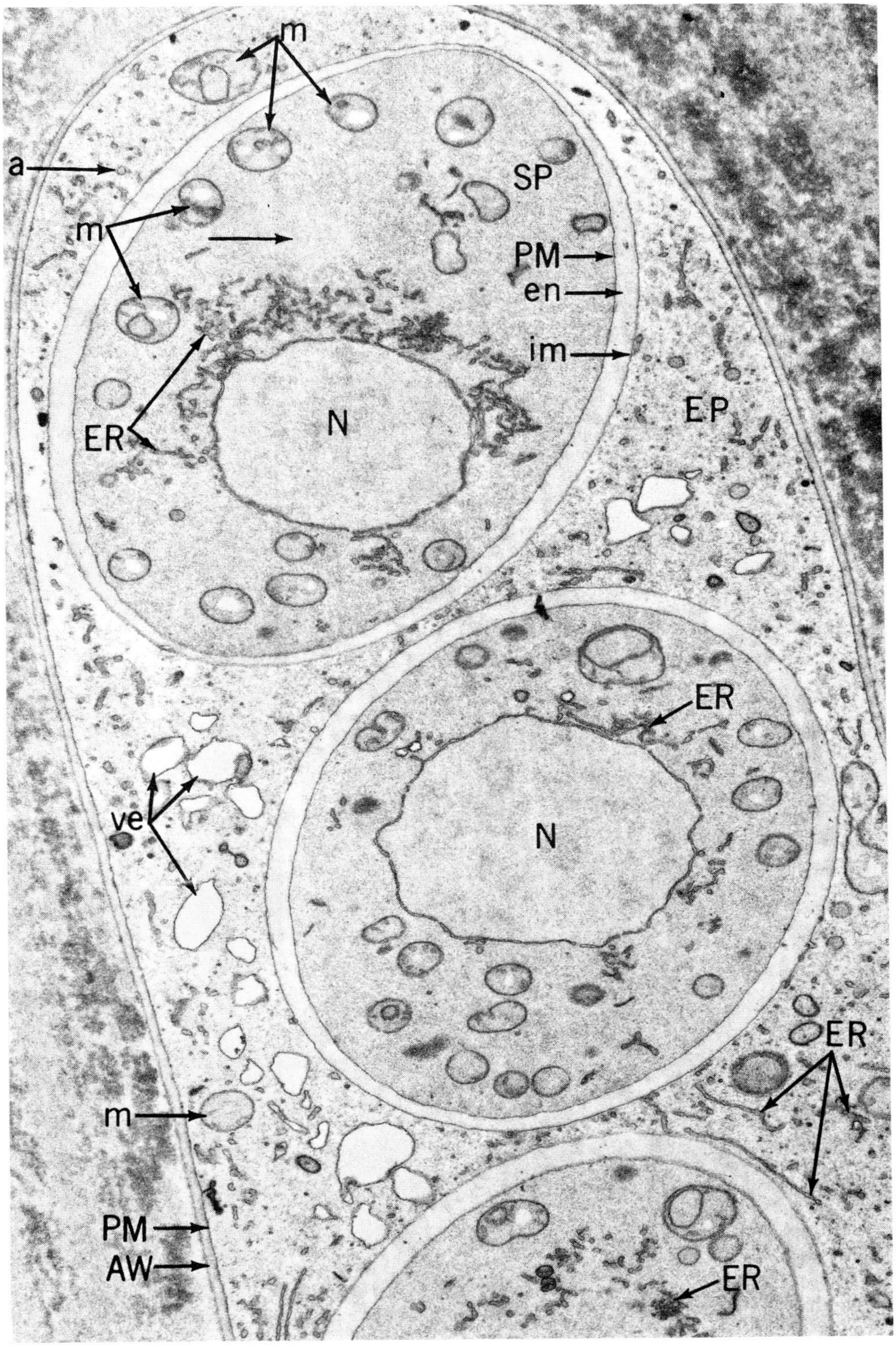
m
a
SP
m
PM
en
im
ER
N
EP
ER
N
ve
ER
m
PM
AW
ER

PLATE 20

Electron micrograph of longitudinal section of an ascus with developing ascospores. The upper ascospore was sectioned longitudinally, and the two median ascospores are seen in cross sections. Fixed in 1.5 percent $KMnO_4$ and stained with lead citrate. × 9,550.

AW, ascus wall; en, endospore; ER, endoplasmic reticulum; im, investing membrane; lo, lomasome; m, mitochondrion; N, nucleus; PA, paraphysis; PM, plasma membrane; ve, vacuole in epiplasm. Unlabelled arrow indicates region that probably contained fibrous or fine-grained structures, both of which are preserved in glutaralydehyde-OsO_4 (see pl. 27, A, B).

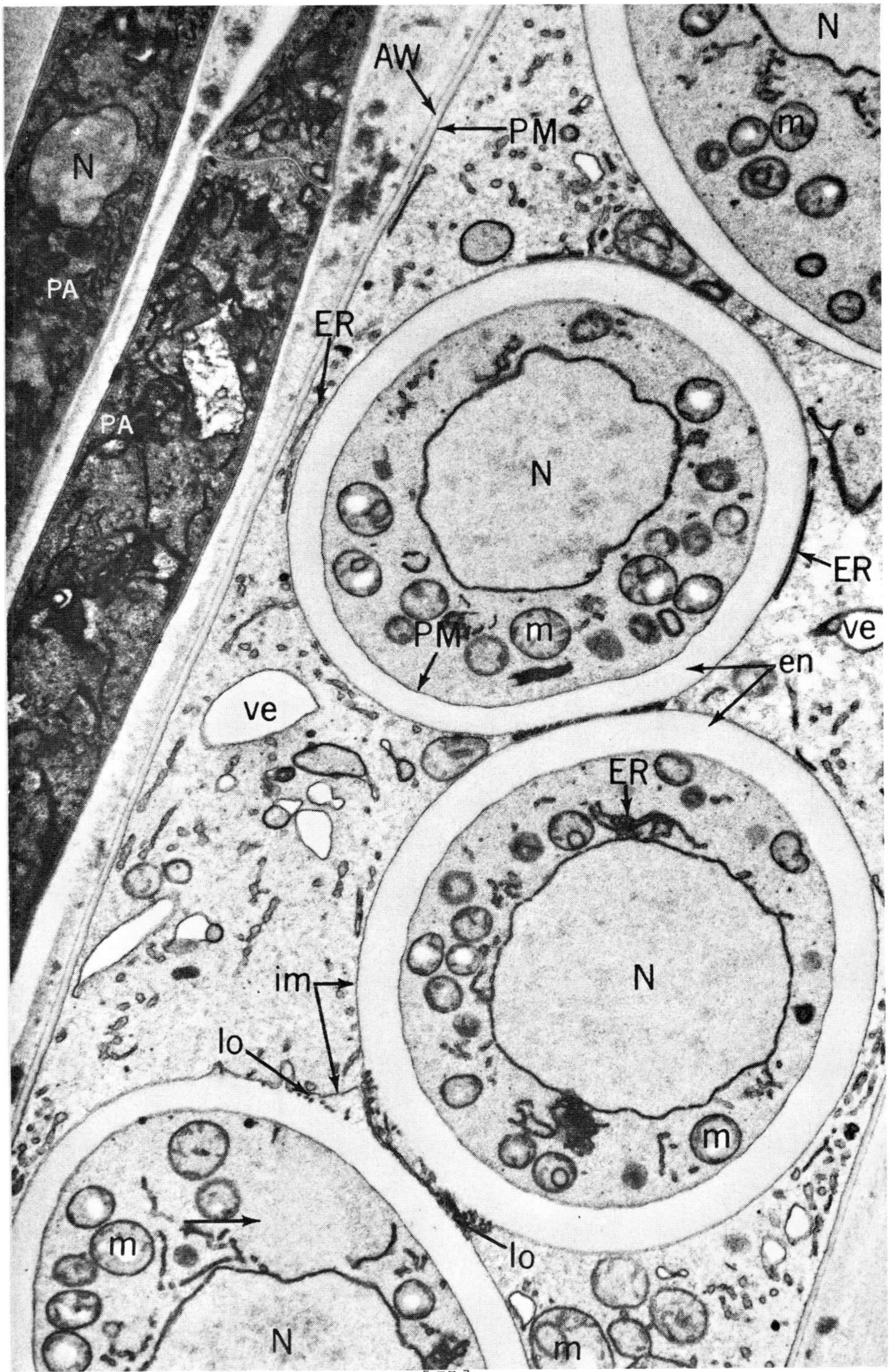
AW
PM
N
m
PA
PA
N
ER
N
ER
ve
PM
m
en
ve
ER
N
im
lo
m
m
lo
N
m

PLATE 21

Electron micrographs of longitudinal sections of asci containing ascospores with developing walls. Fixed in 1.5 percent $KMnO_4$ and stained with lead citrate.

A. Portion of the median sporogenous region of an ascus. × 8,350.

B. Section of apical portion of an ascus. × 9,500.

AW, ascus wall; en, endospore; EP, epiplasm; ER, endoplasmic reticulum; g, glycogen; im, investing membrane; m, mitochondrion; N, nucleus; OP, operculum; PA, paraphysis; PM, plasma membrane; SP, sporoplasm; V, vacuole; ve, vacuole in epiplasm; WB, Woronin body. Unlabelled arrow indicates region that probably contained fibrous or fine-grained structures, both of which are preserved in glutaraldehyde-OsO_4 (see pl. 27, A, B).

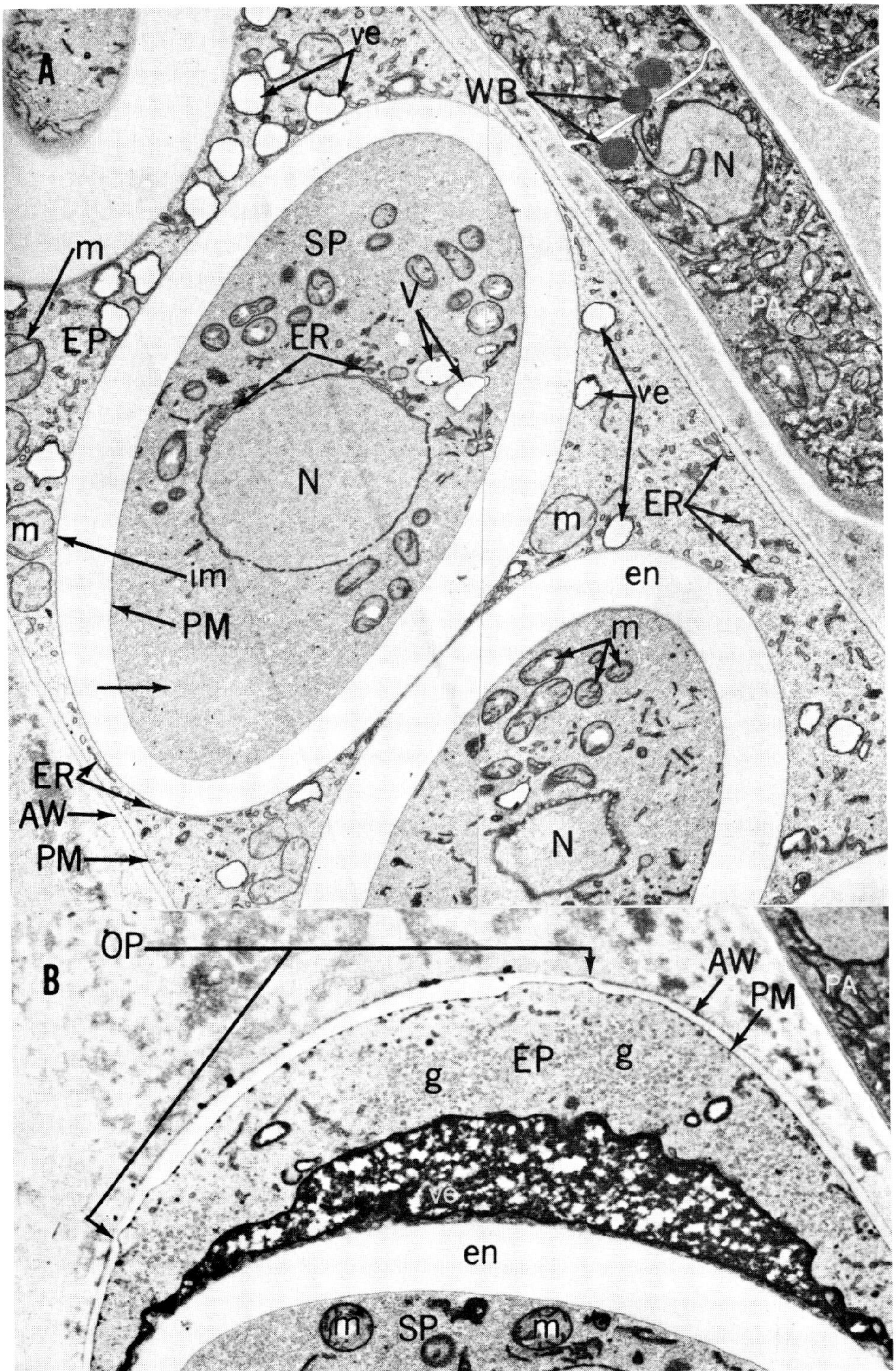
A
ve
WB
N
SP
m
EP
V
ER
ve
ER
m
N
m
im
en
PM
m
ER
AW
PM
N
PA
B
OP
AW
PM
PA
EP
g
g
ve
en
m
SP
m

PLATE 22

Electron micrographs of longitudinal sections of portions of asci containing ascospores with the endospore deposited.

A. Section illustrating a portion of the ascospore wall and adjacent epiplasm and ascus wall. Fixed in 1.5 percent $KMnO_4$ and stained with lead citrate. × 26,800.

B. Section showing portions of endospores of two ascospores and adjacent epiplasm. Fixed in 6.5 percent glutaraldehyde and 1 percent OsO_4 and stained with lead citrate. × 23,250.

AV, autophagic vacuole; AW, ascus wall; en, endospore; EP, epiplasm; ER, endoplasmic reticulum; g, glycogen; im, investing membrane; m, mitochondrion; N, nucleus; pl, perisporal sac; PM, plasma membrane; T, tonoplast; ve, vacuole in epiplasm.

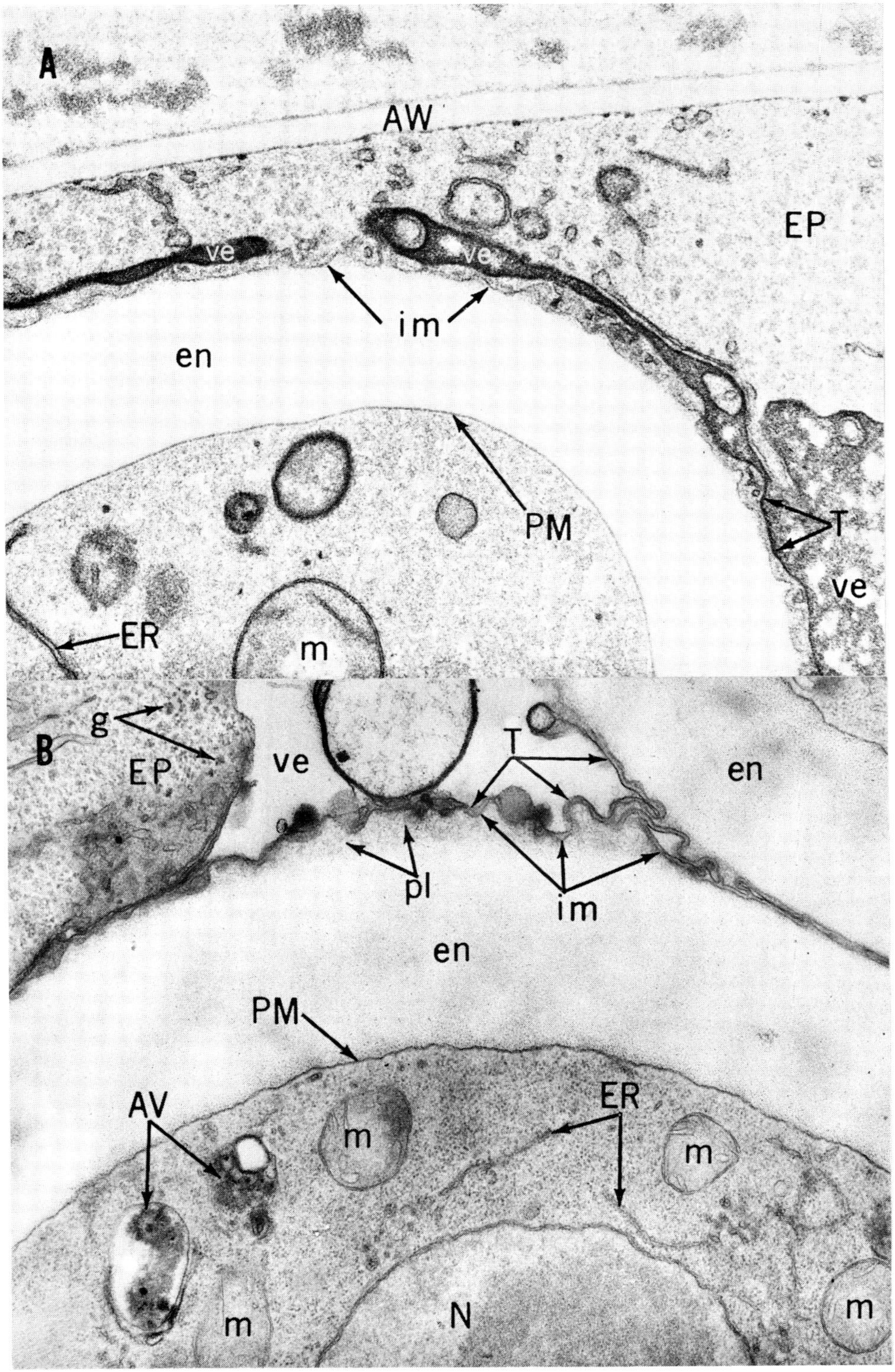
A
AW
EP
ve
ve
im
en
PM
T
ve
ER
m
B
g
EP
ve
T
en
pl
im
en
PM
AV
m
ER
m
m
N
m

PLATE 23

Electron micrograph of longitudinal section of an ascus with developing ascospores. The upper ascospore was sectioned nearly longitudinally. Fixed in 1.5 percent $KMnO_4$ and stained with lead citrate. $\times$ 9,500.

AW, ascus wall; ce, cytoplasm of epiplasm; en, endospore; ER, endoplasmic reticulum; g, glycogen; im, investing membrane; m, mitochondrion; N, nucleus; PM, plasma membrane; T, tonoplast; ve, vacuole in epiplasm.

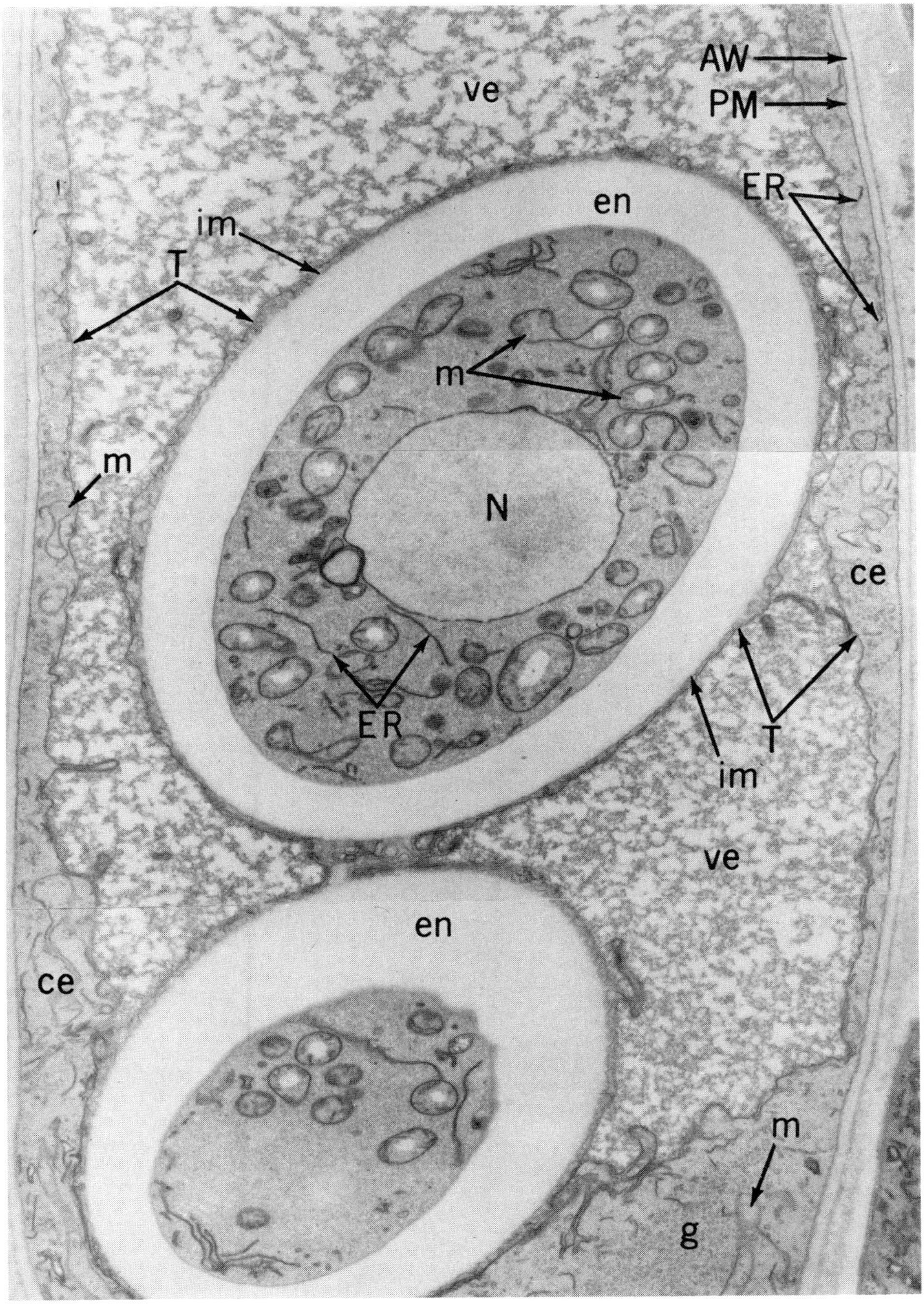
AW
PM
ve
en
ER
im
T
m
m
N
ce
ER
T
im
ve
en
ce
m
g

PLATE 24

Electron micrographs of developing ascospores.

A. Longitudinal section of a portion of an ascus containing a spore with developing wall and perisporal sac. Fixed in 1.5 percent $KMnO_4$ and stained with lead citrate. × 9,000.

B. Section of an ascospore with nearly mature endospore and epispore and pigmented perisporal sac. Fixed in 1 percent glutaraldehyde and 2 percent OsO_4 and stained with barium permanganate, lead citrate, and uranyl acetate. × 8,300.

AW, ascus wall; ce, cytoplasm of epiplasm; en, endospore; epi, epispore; ER, endoplasmic reticulum; g, glycogen; im, investing membrane; L, lipid globule; m, mitochondrion; N, nucleus; PGR, pigment granules in perisporal sac; pl, perisporal sac; PM, plasma membrane; T, tonoplast; ve, vacuole in epiplasm.

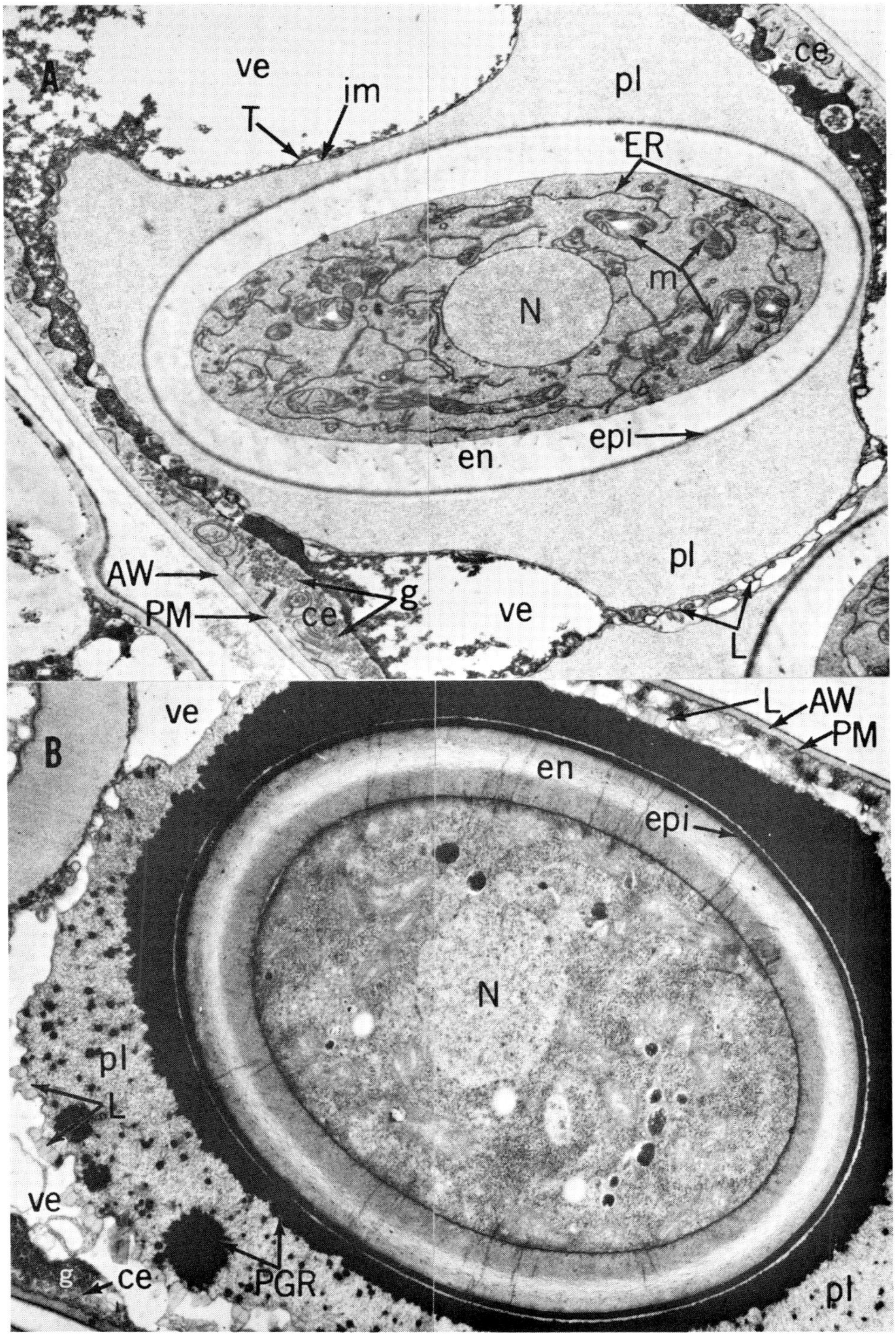
A
ve
T
im
pl
ce
ER
m
N
epi
en
pl
AW
PM
g
ce
ve
L
B
ve
L
AW
PM
en
epi
N
pl
L
ve
g
ce
PGR
pl

PLATE 25

Electron micrographs illustrating the developing epispore and a nearly mature operculum.

A. Segments of three ascospores with developing epispores. The endospore and epispore of the middle spore are sectioned obliquely. Fixed in 1 percent glutaraldehyde and 2 percent OsO_4 and stained with barium permanganate, lead citrate, and uranyl acetate. × 15,200.

B. Segment of an ascospore with a developing epispore and a pigmented perisporal sac. Fixed in 1 percent glutaraldehyde and 2 percent OsO_4 and stained with barium permanganate, lead citrate, and uranyl acetate. × 34,000.

C. Apex of an ascus containing ascospores with developing epispores. Fixed in 6.5 percent glutaraldehyde and 1 percent OsO_4 and stained with barium permanganate. × 11,250.

en, endospore; epi, epispore; g, glycogen; L, lipid globule; OP, operculum; PGR, pigment granules in perisporal sac; pl, perisporal sac; SP, sporoplasm; ve, vacuole in epiplasm.

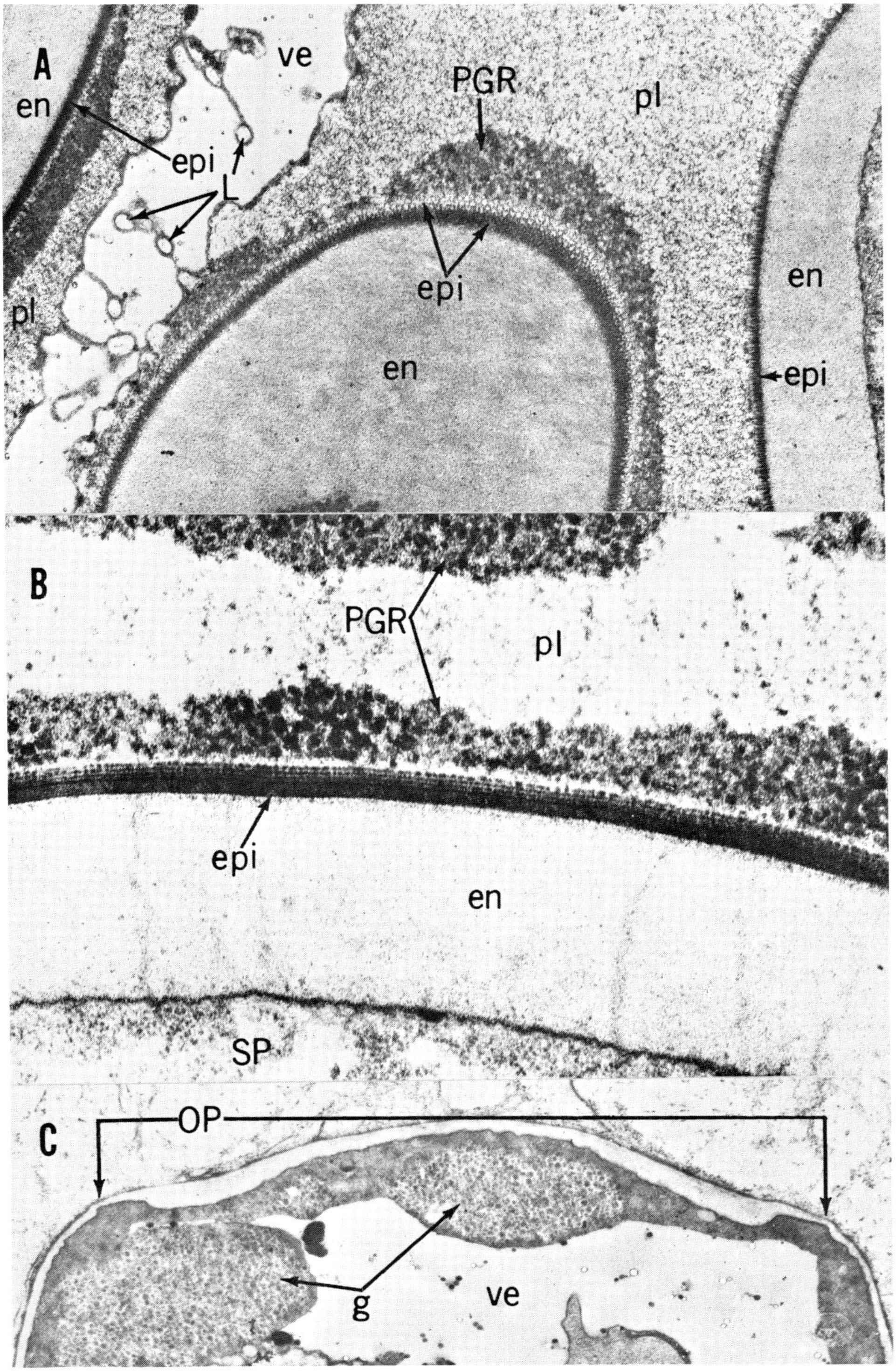
A
en
ve
PGR
pl
epi
L
epi
pl
en
en
epi
B
PGR
pl
epi
en
SP
C
OP
g
ve

PLATE 26

Electron micrographs of portions of nearly mature ascospores. Fixed in 1 percent glutaraldehyde and 2 percent OsO_4 and stained with barium permanganate and lead citrate.

A. Portion of an ascospore wall illustrating the mature endospore and epispore with the adjacent perisporal sac. × 50,650.

B. Portion of an ascus showing the developing perisporal sac of two adjacent ascospores. × 38,000.

en, endospore; epi, epispore; im, investing membrane; L, lipid globule; PGR, pigment granules in perisporal sac; pl, perisporal sac; SP, sporoplasm; T, tonoplast; ve, vacuole in epiplasm.

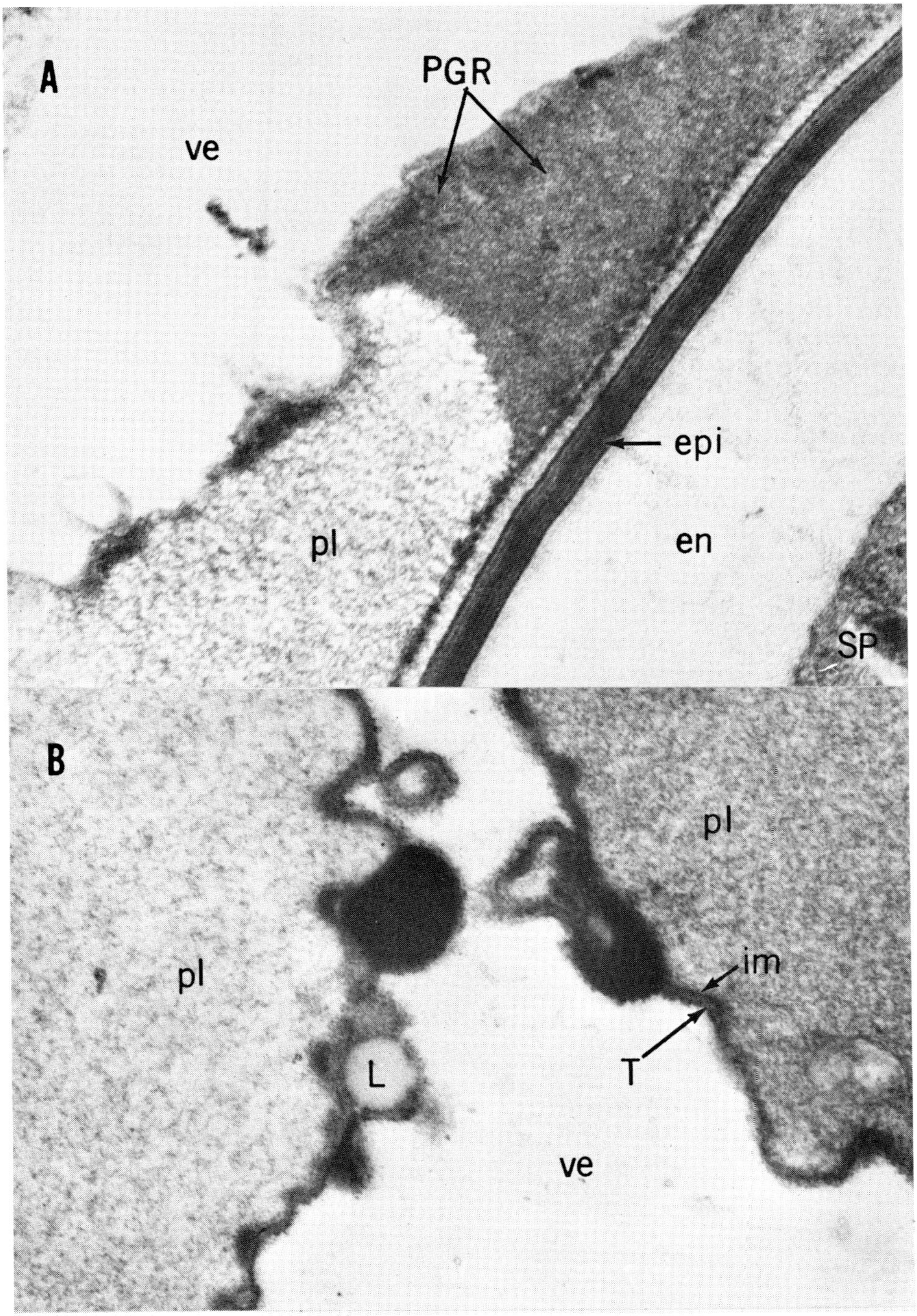
A
PGR
ve
epi
pl
en
SP
B
pl
pl
im
T
L
ve

PLATE 27

Electron micrographs of ascospores fixed in 1 percent (A) or 6 percent (B) glutaraldehyde and 2 percent OsO_4 and stained with barium permanganate, lead citrate, and uranyl acetate.

A. Portion of the sporoplasm of an ascospore with the endospore formed. × 24,000.

B. Portion of the sporoplasm of an ascospore at essentially the same stage as the spore in A. × 49,000.

AW, ascus wall; en, endospore; PM plasma membrane; ve, vacuole in epiplasm.

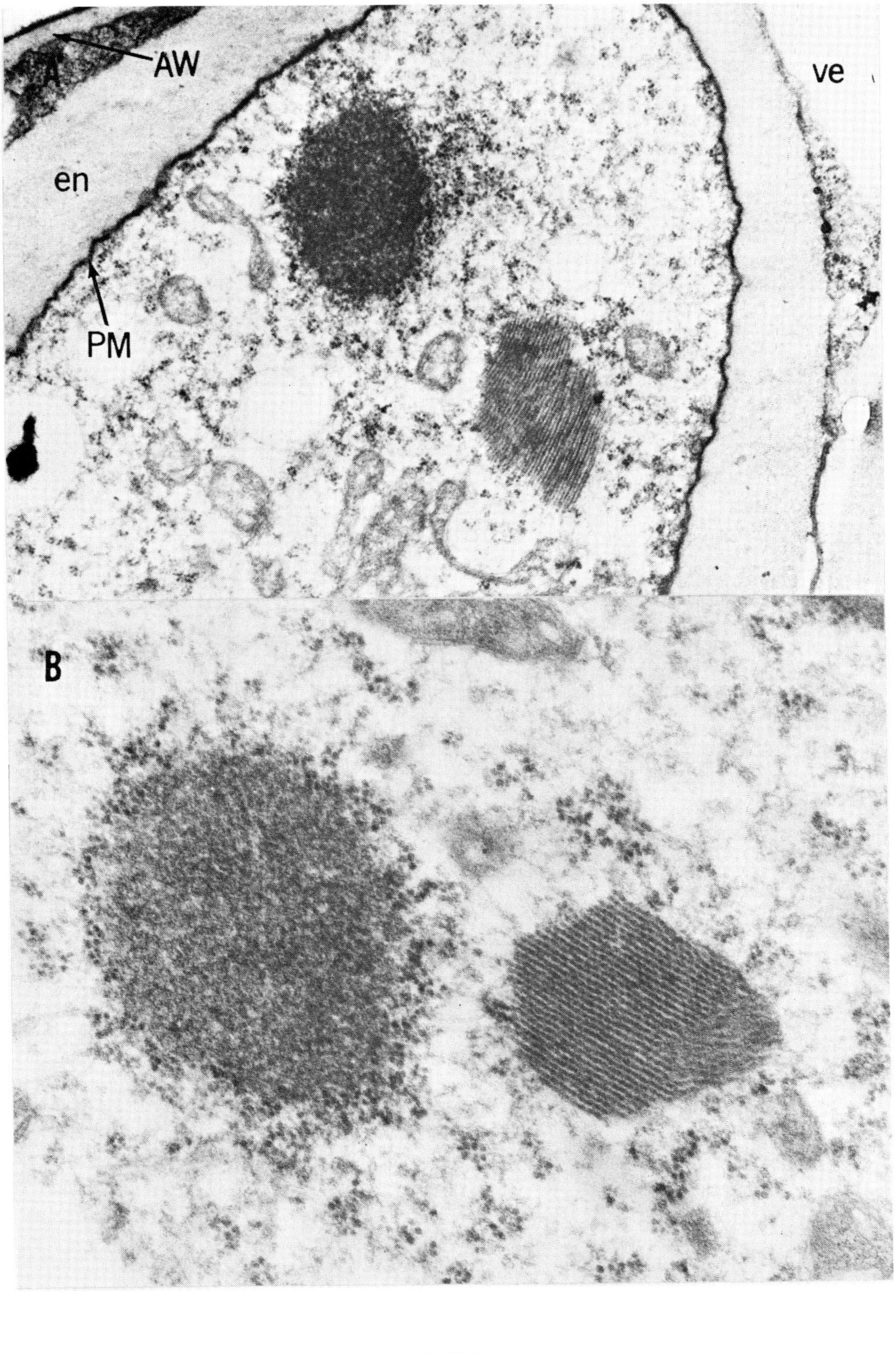
A
AW
ve
en
PM
B

PLATE 28

Electron micrographs of segments of asci illustrating the concentric membrane complex. Fixed in 1 percent glutaraldehyde and 2 percent OsO_4 and stained with barium permanganate, lead citrate, and uranyl acetate.

A. Portion of the cytoplasm of the sporogenous zone adjacent to the wall of a four-nucleate ascus with developing double ascospore-delimiting membranes. × 63,000.

B. Portion of the cytoplasm adjacent to the wall of an ascus with a diploid nucleus. × 43,650.

C. Portion of the cytoplasm of an ascus containing a diploid nucleus. × 44,450.

D. Portion of the sporogenous zone of an eight-nucleate ascus with inwardly developing double ascospore-delimiting membranes. × 43,200.

AD, double ascospore-delimiting membranes; AW, ascus wall; CM, concentric membrane complex; g, glycogen; N, nucleus; PM, plasma membrane.

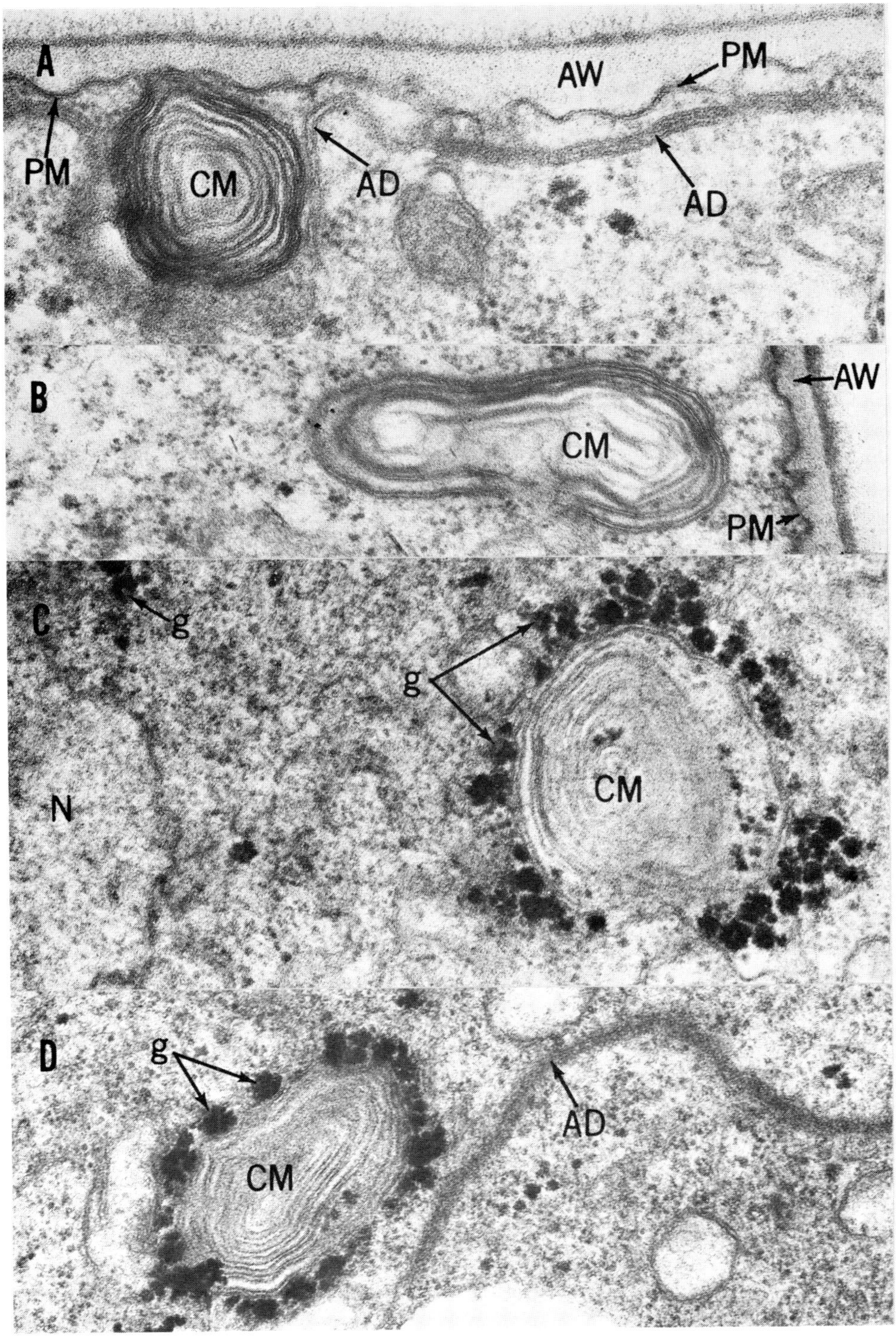
A
AW
PM
PM
CM
AD
AD
B
AW
CM
PM
C
g
g
N
CM
D
g
AD
CM

PLATE 29

Electron micrographs of segments of asci showing various configurations formed by the invagination of the plasma membrane.

A. Peripheral portion of an ascus containing a diploid nucleus. Fixed in 1 percent glutaraldehyde and 2 percent OsO_4 and stained with uranyl acetate and lead citrate. × 36,800.

B. Peripheral portion of an ascus containing a diploid nucleus. Fixed in 1 percent glutaraldehyde and 2 percent OsO_4 and stained with uranyl acetate and lead citrate. × 34,800.

C. Peripheral portion of an ascus at the four-nucleate stage. Fixed in 6 percent glutaraldehyde and 2 percent OsO_4 and stained with uranyl acetate and lead citrate. × 29,000.

AD, double ascospore-delimiting membranes; AW, ascus wall; g, glycogen; m, mitochondrion; PM, plasma membrane.

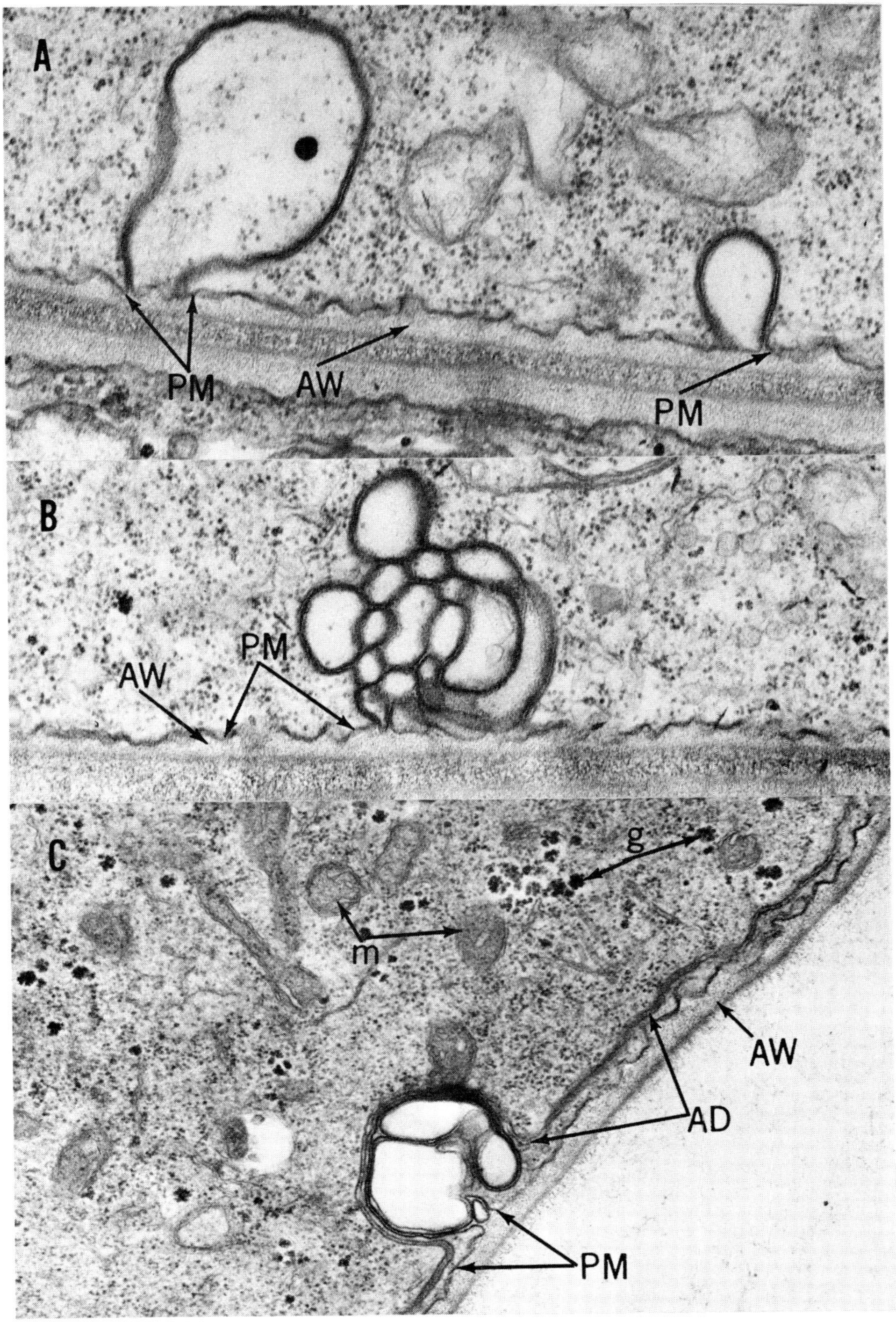
A
PM
AW
PM
B
PM
AW
C
g
m
AW
AD
PM